TRUE GIANTS

Books by Mark A. Hall

Merbeings: The True History of Mermaids and Lizardmen (forthcoming)
Thunderbirds: America's Living Legends of Giant Birds
Living Fossils: The Survival of Homo gardarensis, Neandertal Man, and
  Homo erectus
The Yeti, Bigfoot & True Giants
Natural Mysteries: Monster Lizards, English Dragons, and Other
  Puzzling Animals

Selected Books by Loren Coleman

Bigfoot in Maine (with Michelle Souliere) (forthcoming)
Monsters of Massachusetts: Mysterious Creatures in the Bay State
  (forthcoming)
Mothman: Evil Incarnate (forthcoming)
The Field Guide to Bigfoot and Other Mystery Primates (with Patrick
  Huyghe)
The Field Guide to Lake Monsters, Sea Serpents and Other Mysterious
  Denizens of the Deep (with Patrick Huyghe)
Bigfoot! The True Story of Apes in America
Mysterious America
Weird Virginia (with Jeff Bahr and Troy Taylor)
Weird Ohio (with James Willis and Andy Henderson)
Monsters of New Jersey: Mysterious Creatures in the Garden State (with
  Bruce Hallenbeck)
Mothman and Other Curious Encounters
Cryptozoology A to Z (with Jerome Clark)
Tom Slick: True Life Encounters in Cryptozoology
Tom Slick and the Search for the Yeti
Curious Encounters
The Copycat Effect
Suicide Clusters
Creatures of the Outer Edge (with Jerome Clark)
The Unidentified (with Jerome Clark)

TRUE GIANTS

Is *Gigantopithecus* Still Alive?

Mark A. Hall
and
Loren Coleman

ANOMALIST BOOKS
San Antonio * New York

Portions of this book are based, in part, on previously published material by Mark A. Hall in *The Yeti, Bigfoot & True Giants*; by Hall in past issues of *Wonders*; by Loren Coleman and Patrick Huyghe in *The Field Guide to Bigfoot and Other Mystery Primates* (Anomalist Books, 2006); and by Coleman in his Cryptomundo postings, 2006-2010.

Appendix D: Giant Skulls is excerpted from *More "Things"* by Ivan T. Sanderson. New York: Pyramid Books, 1969.

Appendix E: The Toonijuk is excerpted from *"Things"* by Ivan T. Sanderson. New York: Pyramid Books, 1967

Appendix F: The Teeth of the Dragon is reprinted with permission: Pettifor, Eric. 1995. "From the Teeth of the Dragon: Gigantopithecus blacki." In *Selected Readings in Physical Anthropology*. 2000. pp 143-149. Peggy Scully, Ph.D., Editor. Kendall/Hunt Publishing Company.

Cover art © Alika Lindbergh/Musée de Zoologie Lausanne/ AgenceMartienne

Book design by Seale Studios

For information, go to anomalistbooks.com, or write to: Anomalist Books, 5150 Broadway #108, San Antonio, TX 78209

TABLE OF CONTENTS

To Ivan T. Sanderson and Bernard Heuvelmans,
the Godfathers of Cryptozoology

"By the late 1960s, some researchers began to realize that something bigger than Bigfoot was out there being seen and leaving enormous tracks nearly 2 feet long. One of these researchers, the Canadian John Green, had noticed from the accounts he had collected in North America that a whole group of 'giants' existed who were clearly bigger than the Sasquatch of the Pacific Northwest. Green was convinced that the evidence supported their existence, because he had talked to the witnesses who were very certain as to what they had seen. But it was researcher Mark A. Hall who first gave this group of creatures the name 'True Giants.' Hall had spent years examining the growing body of data pointing to this distinctive group of extremely large hairy hominids who routinely left long, four-toed footprints."

—Loren Coleman and Patrick Huyghe in *The Field Guide to Bigfoot and Other Mystery Primates* (HarperCollins 1999; Anomalist Books 2006).

PREFACE
THE STORY BEHIND THE COVER

CRYPTOZOOLOGIST BERNARD HEUVELMANS died on August 22, 2001, at the age of 84 at his home in Le Vesinet, France. His former wife, colleague, and artist collaborator, Alika (Monique Watteau) Lindbergh, had cared for him in his declining years. During his life, Heuvelmans journeyed without fanfare from the shores of Loch Ness to the jungles of Malaysia, from Africa to Indonesia, interviewing witnesses and examining the evidence for cryptids. As fate would have it, he conducted his last fieldwork before his death by exploring Johor, Malaysia, in April 1993, searching for evidence of the incredibly tall hairy hominoids he had heard about from the area. In this book, we call these creatures True Giants.

Some remarkable interviews came out of this last field investigation by Heuvelmans. One of the interviews resulted in an eyewitness account that the painter Alika Lindbergh, niece of the famous pilot of the "Spirit of Saint Louis," the airplane that first flew non-stop over the Atlantic, would bring to life. She has given us permission to use her illustration for our cover, which depicts the "supergorilla," said to be nine feet (three meters) tall, which inhabits the rainforests of Johor and Pahang in Peninsular Malaysia. Heuvelmans met people who had seen this animal—and had even touched it.

Heuvelmans had decided to spend three months exploring the jungles of the western Malaysia to verify all the accounts of these wild and hairy men of the region he had accumulated during an exchange of correspondence that had lasted two years. He felt

1

there were gigantic anthropoids still unknown to zoology living in the area. He identified them as most probably *Gigantopithecus*, the greatest known primate, which lived around half million years ago in southern China and became extinct 100,000 years ago.

One story especially fascinated Heuvelmans, and that's Captain Mokhtar's story, which is behind the Lindbergh art that appears on the cover of this book. While staying at the park at Lake Cini, Heuvelmans was introduced to, and interviewed, Captain Mokhtar Mohamad, 46, who was the manager of the guesthouses there. Heuvelmans found him to be a cultured man, an experienced expert of the jungle, who spoke perfect English and who was absolutely prosaic in his convictions. Mokhtar, for example, did not believe in the existence of the lake monsters of the Tasik Cini, which is often compared to the Monsters of Loch Ness.

Mokhtar, likewise, was skeptical of a specific serpent-bound superstition of Thai/Chinese origin, which had reached the area with the succession of immigrants from the north. This superstition was fed by the presence in the region of gigantic pythons. The memory of the killing of one of them was still alive when Heuvelmans visited. It was said to be as great as a tree, and issued forth from a bomoh (wizard). But Mokhtar wasn't buying it.

But when it came to the reports of the Orang Dalam, Mokhtar had a different reaction. He had encountered this hairy manlike creature and knew that it was real.

Mokhtar happened upon the creature in the spring of 1989. At almost 6 o'clock in the afternoon, Mokhtar was walking together with his wife, Azlinah Ismaïl, along a footpath near the jungle at Palau Balaï, an islet on Lake Cini. He soon heard a loud noise and sought its origin. He discovered that the sound was produced by a big animal advancing towards them in the foliage.

Suddenly, Mokhtar felt a large hairy hand on his left shoulder. Reacting instinctively, he struck at the creature with his parang. (The parang is the Malay equivalent of the machete. Typical vegetation in Malaysia is more woody than in South America, and the parang, with its heavier blade and a "sweet spot" further forward of the handle, is therefore optimized for a stronger chopping action.)

As soon as Mokhtar struck the creature's hand, he almost immediately lamented what he had done. He would tell Heuvelmans that he would regret for his whole life this violent but, at the time, comprehensible gesture. As the creature ran away, Captain Mokhtar noticed it did so in an erect position, racing along and making oscillations with his arms like a gibbon. There was no indication of a tail at the base of the back. In the tracks that the giant ape left behind, the imprint of the heel appeared more deeply marked than that of its four visible toes, a sign of its very particular walk.

From her vantage point, Mokhtar's wife, Azlinah Ismaïl, saw just how much the huge beast dominated over her husband, who measured 6 feet 8 inches tall. The creature was similar to an enormous gorilla, they both said, entirely covered with fairish hair. Its head was bigger than that of a man's, and it had long blonde hair, which fell on its wide and mighty shoulders. Azlinah, then six months' pregnant, was so shaken by the encounter that she aborted.

While some visitors of the park's guesthouses would sometimes report catching a glimpse of these creatures at night, Mokhtar never saw one again.

INTRODUCTION

WE LIVE IN unique times, in which we are discovering that, clearly, we are not alone. In the March 25, 2010, issue of *Nature*, Johannes Krause and his colleagues announced the complete mitochondrial sequence of a pinky bone unearthed from Denisova Cave, in the Altai Mountains of Siberia. That work goes far in proving that the source hominin was much different than anything ever recognized. These researchers estimate the age of the little finger (*digitus mínimus mánus* or pinky) to be between 30,000 and 48,000 years old. Therefore, we now know that the Denisova hominid, or the X-Woman as she is also being called, lived at the same time (40,000+/- years b.p.) as humans (*Homo sapiens*), Neandertals (*Homo neanderthalensis*), and hobbits or Flores people (*Homo floresiensis*).

Regarding the new Denisova discovery, Ian Tattersall of the American Museum of Natural History in New York was quoted as observing: "The human family tree has got a lot of branchings. It's entirely plausible there are a lot of branches out there we don't know about" [1].

And Terence Brown of the University of Manchester stated: "Forty thousand years ago, the planet was more crowded than we thought" [2].

In 2003-2004, when the discovery of the hobbits was announced (between nine and eleven *Homo floresiensis* have now been found), the tales and folklore of little *ebu gogo* from the island of Flores were gathered and said to relate to the find of the little woman of Flores. Now the world is being introduced to the "X-Woman," which actually seems to be a young female.

5

We expect there is a link between the stories of Wildpeople of southern Siberia — namely the Almas, Chuchunaa, and Mulen — and this new X-Woman population.

So the question that naturally arises, given all these newly recognized branching of the human family tree, is: Did Giants once roam this Earth, too? Yes, we know they did. The fossil record does not lie. Do they still exist today? You can be the judge of that. But we hope that this collection of tales and accounts will make the legend of the True Giants come to life for you. Remember that these tales of giants are the reflections of authentic encounters.

Skeptics tend to regard cryptozoologists as folklorists. Daniel Loxton, the editor of *Junior Skeptic*, sees the role of cryptozoologists "as being primarily folkloric, with the potential existence of these creatures being a spin-off question from the first task of collecting and archiving these tales" [3].

Indeed, there is no denying that we feel we have captured the myths, legend, and folklore of the True Giants in these pages. And given our research, we also sense there are some facts behind these legends. As the recent reports from Asia show, the presence of the giants in our world very likely extends into the present day. We await the time when the announcement is made that the modern survivors of the fossil primate *Gigantopithecus* do indeed exist.

Chapter One
The Universal Giant

WE HAVE ALL heard of giants. Today, the word is likely to bring to mind the few examples of human beings who have grown to extreme sizes. Robert Wadlow is one who was known as a giant in the 20th century. He reached a height of 8 feet 11.1 inches and is regarded as "the tallest person in history." Typically, such human giants reach heights of less than 9 feet. All claims to greater heights for human beings appear to have been exaggerations.

There have been stories of greater giants, of course. The storybook giants were enormous creatures taller than the houses of past ages and even today's houses, too. As one author noted of them: "They raised terror among sheep and bad little boys." In history the appearances of these enormous men reach back to times when historical records were just beginning to become commonplace. There were still other giants like Goliath in ancient history.

Could there have been a basis in the natural world for such enormous creatures? We will explore here the subject of True Giants: what they were, where they came from, and what they are today. They are the base upon which storybook giants were invented. They are the reason that people in ancient times spoke about famous giants in their world. And they are the reason that people still see tree-tall giants in modern times. Not many people get to see them; the giants spend most of their time in places remote from people. But they still bump into people now and then, leaving behind them appropriately large footprints. Their

Gigantis Sceleton

Ancient traditions speak of the existence of giants. From Mundus Subterraneus *by the German Jesuit Athanasius Kircher, published in 1678.*

tracks can measure from two-dozen inches to three-dozen inches in length. As you can see, we are talking about some truly big fellows.

There is a basis in modern science for discussing the existence of genuine giants of this size. But you may be wondering why—outside of storybooks—you haven't been hearing about them before this.

There are two reasons. First, the existence of True Giants is not a popular idea. Such things are not supposed to be real. So, when people have reported them in places like the Cairngorms in the northern United Kingdom, in Southeast Asia, and in Canada they have been regarded as mistaken or even dishonest tales. Secondly, the fossils that have been found for this particular giant primate have been attributed not to a giant man but, erroneously, to a giant ape. There is no basis in the fossils themselves to support this determination. Rather, it has been merely a popular prejudice among the fossil specialists to make this categorization. Some people have suggested that the fossils, known as *Gigantopithecus*, are gigantic men. We believe that view will one day be proven correct.

Gigantopithecus-sized bones have been found in many places around the globe. But those bones were lost in recent centuries before good notice was taken of them and proper descriptions could be recorded. No one has been running around deliberately looking for gigantic bones in those places. Instead, such bones are still only sought in geological deposits that are a half million years old or older. Some day the finding of more gigantic bones in proper scientific digs will validate a full prehistoric record for True Giants.

But we do know a little something of these giants today. They have been in the news in places like Southeast Asia for decades, and in late 2005 became a news sensation once again. The world was treated to news dispatches from the southern interior of the Malay Peninsula. Everyone was reading about the Orang Dalam (or "Interior Man") of Johor. Gigantic footprints were found and hairy man-like figures were reportedly seen. They were 10 feet tall and more.

This had all happened before, decades earlier. As so often happens, details of those previous events were almost forgotten. The intense interest in the here and now, to be captured in photographs and videos, left it up to those who had been chronicling these happenings for decades to recall the important historical parallels.

The identity of these hairy giants has been obvious for many years, but this too has been overlooked by the reporters focused on the here and now. It was in 1992 that we first connected the specific characteristics of True Giants in modern records with the survival of *Gigantopithecus*, and everything about extremely tall primates that has come to light since then has supported that association.

These True Giants are not "Bigfoot," despite some efforts to make simple comparisons with creatures such as the one seen in the famous Roger Patterson-Robert Gimlin motion film of a Neo-Giant in California in 1967. Those giants get no taller than 9 feet high. They are of a different genus of primate. The creature in the film is an example of how other large and hairy beings are also found in the wild. But they do not grow to the heights of the True Giants. They are different in physical particulars, in their feet (as known from footprints), and in their behaviors.

The tree-tall True Giants were familiar under many names to the original inhabitants of North America. A petroglyph-covered boulder in North Carolina is named the Jutaculla Rock after the legendary giants of the mountains. Table Rock in South Carolina was associated with a chief among the giants.

In eastern Canada, Samuel de Champlain heard about feared giants said to be living in the forests. Other French explorers heard the same and noted how the Indians appeared to be truly frightened by these creatures.

In Alaska True Giants were legendary. One historical record tells how an Indian chief disappeared right after his party saw signs of a True Giant they knew as Gilyuk.

Mountainous regions around the world seem to be populated with True Giant traditions. They suggest that, if those giants are not still around, they were at one time a reality. Sightings of tall figures in mountain mists and thick forests, taken with the occasional record of enormous footprints, all point to the possible survival of True Giants even today.

We are focusing here on those many beings that are the survivors of *Gigantopithecus*, a specific genus of primate that possesses its own lineage among the primates. When carefully examined, they have a physical description that is like no other animal. They possess feet that deposit enormous and unique impressions in the ground. True Giants have their own cultural traits perhaps achieved, in part, by borrowing from others. They have their own behaviors that are a consequence of their unique origins, size, and status among the primates.

All that we know of them now can be drawn from the human records of encounters in the wild, the tracks they have left behind, and the knowledge of ethnic groups around the planet who tell in their local history about these "Big Men."

Let it be clear from the outset that, despite the clear tendency of local ethnic groups to call these tall primates "men," they are most certainly not men (in the sense that they are not *Homo sapiens*), unless we would care to redefine the term to include such a distant primate on the bush of primate evolution. *Gigantopithecus*, we sense, while they may be experienced as "gigantic men" versus "giant apes," nevertheless are not to be defined as "human."

We would say it serves no purpose to do so. The true picture of primate evolution includes many branches with sophisticated capabilities. The recent discovery of *Homo floresiensis* in Indonesia is a good example of what we will be finding in the next decades. These Little People of Indonesia are at the other end of the spectrum of primate diversity. They are dwarfish "hobbits" that are known to have co-existed beside humans as recently as 12,000 years ago. The native people of Indonesia still talk of seeing them alive.

The find of *H. floresiensis* is becoming a paradigm-busting experience for palaeoanthropologists. They are finding this primate had unexpected capabilities, and may have survived into recent times.

There are numerous accomplished primates living today who have remained distant from humans and enigmatic in their appearances. They appear to be capable of culture and language, and it would serve us ill to attempt to fit them all in under an umbrella of "mankind." We cannot stretch a clear definition of "mankind" to encompass the broad field of accomplishments among the higher primates.

Instead, we will have to acknowledge that culture and language are not ours alone. They are characteristics shared among many surviving primate groups: True Giants could be one of those groups.

When first writing about True Giants nearly 20 years ago, Mark Hall referred to them as "the universal giant." One of the surprising things to come to light about True Giants is just how universal they have turned out to be. They are known on small island groups like the Comoros in the Indian Ocean. They have been reported as rafting about the Pacific Ocean. The Eskimos of Greenland say they dwell on the edge of that island's great icecap. The waste areas of Iceland appear to have been inhabited by them in the past.

No one can say precisely where they might still be present today. We have ignored them too well to claim much certainty about their whereabouts. But we can say, based on the records presented in this book, that they may have spread around the globe.

When traditions of such creatures exist even in the Hawaiian Islands, we can no longer be surprised. They have used devices like rafts to travel. And their antiquity also means that their kind might well have moved from land mass to land mass long ago when the geology of the planet was connected with various land bridges and continental configurations.

Their level of success might have varied over time depending upon the competition they had and the development of their own technological skills. More importantly, perhaps, they might have borrowed from others such as human beings who came along later in prehistory.

We are seeing most clearly a snapshot of how these marvelous giants looked in recent centuries. We have hints of how they were, handed down from older traditions among tribes of humans.

We also have specific knowledge of them in fossil finds. Those are limited at present to some areas of Asia. Other finds have suffered the fate of being lost and destroyed from lack of any interest and understanding of what they signified. If we can exercise more discretion when future bones of giants turn up in odd places, we will be able to study them and add to our knowledge about how these giants spread through the last several millions of years to end up where we see them today.

The focus here on True Giants should not be seen as a slight to the Little People of Indonesia (*Homo floresiensis*), the X-Woman of Siberia, the Neo-Giants of California, or any other spectacular surviving relatives of mankind on the tree of primate evolution. They are all of great interest and they are all deserving of our curiosity, time, and careful study. The tasks are enormous, made so by our collective failure to take them seriously and study them in recent centuries as humans have increasingly noticed them.

We are taking a first step here by organizing what we know of one of these fascinating creatures, the True Giant, at the start of the 21st century.

We will review True Giants as they have emerged into human awareness by way of news reports and ethnographic studies, through the collection of folklore, and by certain careful scrutiny of those remains of bones that have not been lost, as too many have been through fear and neglect.

We will discover that True Giants appear to still be present in many places on the planet. We will find they have been seen in places not so far from where many people are living unaware of them.

We will see that this has come about for two main reasons. Firstly, the giants have learned to be secretive, for their history is one of being killed when mankind began to use firearms and other weapons to defend themselves against competitors.

The early conflict between humans and True Giants is illustrated through the story of "Jack and the Beanstalk," which was a variant on the "Jack the Giant-Killer" tales. Jack is shown here presenting the captured wild giant, dressed modestly in oversized clothing, to his king. From The Cruikshank Fairy-Book *(1847), illustrated by George Cruikshank.*

Secondly, we will see that this topic has been an unwelcome consideration in modern times. Our social memory of these creatures has faded rapidly and has been manipulated, psychologically, as well as sociologically. In *Giants in Those Days*, classics professor Walter Stephens, formerly of Dartmouth now at John Hopkins University, argued that the modern view of jovial giants has obscured earlier records that presumed the historical reality of giants and the perception of them as aggressive co-inhabitants of Earth. Writes Stephens:

> A relatively quick experiment in the cultural history of the Giant can be made by observing the evolution since 1878 of the entry 'Giant' in the *Encyclopedia Britannica*. Even well-read people of today may be perplexed to see the foremost general encyclopedia indulging in discussions of 'the conception of giants as special races distinct from mankind'....The same information may be found as late as 1911, except that the *Britannica* then included information on a pioneering study published in 1904 by two French researchers, who had demonstrated the scientific impossibility that Giants ever existed. In fact, Pierre-Emile Launois and Pierre Roy's *Etudes biologiques sur les géants* can be said to have accomplished the scientific demythification of the Giant [1].

People have become less interested in True Giants as they have receded into their secretive ways, especially in a rational climate where the cultural anthropology of our species greatly discourages thoughts in the modern world about dangerous giants. The retiring giants have shown the value of the axiom: "Out of sight, out of mind."

We will bring them to mind here, and start a fresh consideration of the meaning of giant fossils and giant "men" as reported in the 21st century.

Chapter Two
The Earliest True Giant was
Gigantopithecus

Four categories of evidence support the presence of giants as living beings around the world. They have survived into the modern day, despite their limitations in intellect and the enmity of humankind.

The many traditions are one body of evidence to consider. The giants are said to be enormous and fearsome. They are often designated "cannibal giants" because the people saw them as huge counterparts to themselves. The giants appeared to be preying upon their own kind when humans were victimized. The physical aspect of these particular traditional giants has often been left to the imagination of the listener or the reader as stories were passed along through centuries of time. Men who never saw a True Giant for themselves have naturally been the ones who made the artists' renderings.

Another category of evidence suggest that there does exist a consistent appearance for True Giants. Some people have sighted giants and have given us descriptions of what they observed.

A third category of evidence can be found in the size and shape of the tracks left by giants. They have been found where traditional giants could remain and where people have reported them.

And, finally, there are fossil remains that show the origin of True Giants. The first description of these fossils attributed them to *Gigantopithecus,* meaning "Giant Ape." A later opinion suggested they belonged to *Gigantanthropus,* or "Giant Man." We think the latter opinion will one day be proven to be the more accurate one.

We will look at these categories in reverse order, starting with the fossils that are the foundation for the presence of True Giants in our world.

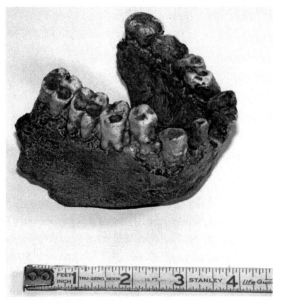

Replica of Gigantopithecus *mandible. Courtesy Lehigh Valley Museum of Natural History.*

BETWEEN 16 AND 10 MILLION years ago a group of primates, the Eurasian Drypopithecines, had evolved. Among them was a man-sized, ape-like creature designated *Dryopithecus.* It appears to have survived to give rise to the Yeti legend of the Himalayas. It had a much larger relative, known as *Gigantopithecus.* This giant is known only from jawbones and teeth [1]. Remains of other parts of the body have not survived in the records found so far.

The fossils are rare. They have been identified in China, India, and Vietnam.

The first bones of *Gigantopithecus* to be recognized were obtained by Gustav Heinrich Ralph von Koenigswald (1902-1982). This German scientist cultivated the habit of examining the contents of apothecary shops in Java and then in China. In this way he was able to spot fossilized bones that had been collected and turned over to druggists. Those specialists in healing arts would grind them up to create potions. Some people had notions that these were "dragon bones." Based only upon his earliest finds of giant teeth, Von Koenigswald suggested that they came from a large ape [2].

His colleague, Franz Weidenreich (1873-1948), was the one who re-examined the giant teeth and subsequent finds. He came to a different conclusion. He thought they were evidence of a giant man and should be called *Gigantanthropus* [3]. The earlier view has prevailed among most scientists to the present day, so the name used in all discussions has remained *Gigantopithecus*.

The best way to appreciate the tremendous size of these creatures might be to compare the fossil jawbone with a modern human jawbone.

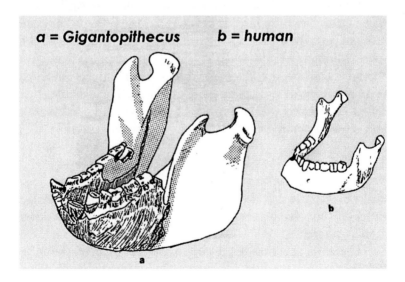

a = Gigantopithecus   b = human

Reconstructions of this animal have depicted it as an over-sized gorilla. One model depicted an upright *Gigantopithecus* to have a height of 10 feet, although we think that mockup is too short [4]. This animal was upright most of the time and was unlike a gorilla in most respects. We think it has reached the heights attributed to the True Giants, exceeding 10 feet. They could reach as much as twice that height.

Estimations are involved in all but one of these cases. A dead True Giant was once measured to be 13 feet tall (that account will be given in detail later). Thirteen feet alone is a staggering contrast for a human being to appreciate. Heights of 15 and 20 feet have been reported after sightings.

What makes these soaring attributes seem possible is that the reported sizes of footprints have also increased accordingly. As reports have told of bigger giants, so have the measurements of the footprints enlarged.

We would allow that once we get to heights of 20 feet we might have reached a size so great that humans would have a hard time reaching a good estimate. The factor of unintended exaggeration might well enter the picture and account for the more spectacular reports.

However, we would not want to preclude the possibility of these giants reaching heights beyond 20 feet. Perhaps there are cases of long-lived True Giants who have grown to exceed the more common heights of 10 to 20 feet among their kind. We should not assume that all observers had a poor frame of reference or were incapable of making good estimates. The extremes that might be attained will be a subject for long-term considerations and is not something that will be resolved any time soon.

PEOPLE ARE REPORTING True Giants in North America. This raises objections that there are no fossil remains found in the Americas, and that a possible migration into the New World was seen as difficult.

There are accounts of gigantic bones having been found in the New World, however. They were simply poorly handled and lost. Scientists were not looking for such evidence and took no interest in the finds when they had a chance to examine them. The history of paleoanthropology in North America has been one of looking for fossil finds within limited parameters of time and species. To stray from those parameters was not "acceptable," and thus reports of "giant bones" were often excluded from consideration. (See Appendix C: Giant Bones.)

With an increased vigilance, we can hope that fossil remains will one day be recognized and preserved.

As for the migration issue, there have been major geological changes on the planet such as are recognized in the Pleistocene. True Giants could have walked from the Old World to the New when a land bridge permitted it. Or they might have rafted their way, as one Latin American tradition has described.

As human beings made the transfer from the Old World to the New, these other primates likely did the same, and we can speculate that they might have done so before humans made the move.

Observations of True Giants in modern times have not been frequent, but we cannot be sure how often they might have been seen. If, in your headlights one night, you saw a figure 20 feet high cross the road in front of you, how many people would you tell? Probably no one.

There is a lot recorded in American Indian lore for True Giants. They have left their mark in the geographical names given to places in North America. And there are modern reports for them as well.

They are around in many parts of the world. To survive at all they remain shy of human beings. It is no accident that the detailed observations of these giants are so often made from a distance and that the best records come from mountainous areas as they do in North America.

First of all, the surviving giants no longer confront human beings if it can be avoided. In rare instances of prolonged visual contact they have kept their distance from observers. Secondly, in the New World's Pacific Northwest there has been an unparalleled effort to collect accounts of hairy beings of all kinds. If comparable efforts were made elsewhere, we would be likely to hear of similar matter-of-fact and detailed sightings of True Giants.

Before unwarranted assumptions about the body and the mass of True Giants are made, the many unknowns should be stressed. We don't know how much of the impressive size of True Giants is only the hairy exterior and what the body beneath the hair is like. We don't know how efficient this body is, how graceful or awkward it is, what ills it is prone to, and how active or inactive it must be in any one day. The unanswered questions about True Giants are legion, such as the extent of their culture, their speech and language, if any, and even the specifics of their diet.

A comment here about their diet: Their reputation as "cannibals" (meaning in folklore they ate humans) should not cloud our perceptions of their diet. To support their size the giants are probably omnivorous, meaning they eat plants and animals. The teeth of *Gigantopithecus* are considered to have been adapted to graminivorous feeding, such as grinding roots and grass stems [5]. The suggestion here is that they also added meat to their diet, if not at an early date, then certainly in recent centuries.

The modern appearances of True Giants are distinguished most often by the unique tracks they leave. Those tracks are enormous and show only four prominent toes. There is sometimes the trace of a fifth toe on the inside of the foot. These giants almost certainly have a mammalian limb so the bones for five digits are accounted for in some way in the course of their evolution.

The feet of True Giants support a great height and it may be that the hallux (normally a big toe) provides added assistance in balancing their upright stance; it appears to be a smaller toe than the rest and to serve in some auxiliary capacity.

People convinced that they know how "genuine Bigfoot tracks" should look have dismissed such tracks. Large five-toed tracks also exist that can be attributed to tall creatures known as the Neo-Giants. Those Neo-Giants get no taller than 9 feet in height. True Giants are something different and their tracks typically show only four prominent toes. The True Giants are also distinguished by their behavior, their physical appearance, and their origins in primate evolution.

There is an old saying that "the map is not the same thing as the land it shows." We should keep in mind also that "the track is not the foot." An impression left in a medium of sand, soil, snow, etc. does not show us the structure of the foot. It is merely a disturbance made by the motion of the foot. Impressions can be incomplete, as when running or walking on tiptoes. Four-toed tracks are another case altogether. The impression shows us the result of the evolution of the primate foot to carry the size of True Giants. The foot is flat and broad, and sometimes has only four *apparent* toes of similar size. In width it is one-half of the total length of the foot. The fifth toe has become vestigial and is not prominent or might even be absent in some tracks.

The same tendency toward a broad foot with fewer digits is apparent in the Yeti track. The Yeti is a squat and bulky creature that can be associated with Dryopithecine origins. It has a broad track that has nearly evolved to the point of appearing to have only three toes. This is because three of the toes have become so small and bunched that they appear to be almost one toe. There was early confusion about the number of toes on this type of track when it was first described in Asia.

The best records of True Giant tracks have been preserved in North America. Examples have been reported at Cold Lake, Alberta (10 x 21), in June 1976; Snoqualmie, Washington (8 x 17), in January 1976; and outside Astoria, Oregon (7.5 x 17), in December 1977.

A graph of the widths and the lengths of these tracks show a consistent average slope line running from a point of 3.5 x 9 inches to the point of 12 x 24 inches.

There has been no direct evidence to link *Gigantopithecus* with the four-toed tracks and the True Giants of folklore. The large imprints with four big toes and one toe on the side have been found and associated with the Orang Dalam. Those giants look like a living embodiment of *Gigantopithecus*.

We have had the best record for True Giants in modern times in North America because more effort has been put forth there to gather the records of encounters. But now that record could be eclipsed by close studies of the giants in Southeast Asia.

The surviving descendants of *Gigantopithecus* have been identified in Europe, Asia, and North America as upright and lean near-men of spectacular height. Those who wish to see the proof of giants need only examine the jawbones and teeth of this fossil type. They are the only remains that have been revealed by scientists. Many other alleged bones and skulls of giants could be mentioned, but they were found centuries before the invention of the physical anthropologist. They were not preserved.

We can explain the widespread occurrence of giants. Giants have been interpreted as personifications of natural phenomena, as older gods in conflict with newer ones, and as demons from a realm of the dead. Regarding another view of giants, that they were some kind of giant man, mythologist John A. MacCulloch has this to say:

> They have been regarded as an earlier and wilder race of men, with stone weapons, opposed to the more cultured race which uses the plough... The wilder traits of giants suggest a savage race, but the theory does not explain the universal belief in giants nor the great stature ascribed to them [6].

The stature of giants is explained by their identity as a different genus of primate, i.e., *Gigantopithecus*. The "universal" nature of giants could be explained by the following hypothesis.

The distribution of "giant men" is nearly coincident with the dispersal of human beings around the globe. Our hypothesis to explain the success of True Giants would be that they were good at mimicking the cultures of human beings. In clothing, language, and subsistence patterns they may have done well by copying the models they observed in neighboring humans.

The brains of these giants combined with their physical skills give them the capacity to live like human beings, communicate with humans, and adapt to the world's varied climates from the frigid north to the equatorial regions. The giants were successful in ways that human beings demonstrated first.

The surviving knowledge of True Giants suggests to us three

phases of their interaction with human beings. The first phase tells us of ancient times when giants lived like humans and shared the language and culture of the smaller and weaker men and women. The giants' abilities allowed them to spread across the globe, as did their smaller relatives. The extent of giant penetration into Africa and South America is still problematic, although this situation may only reflect our collective inattention to the subject in those regions.

The next phase is that of a schism between the species. True Giants may have been too successful at the art of mimicry. With their superior size, they would have been, after all, always somewhat menacing. They had great appetites and were competing with humans for the same food and living space. They were notorious for resorting to cannibalism. This particular "cannibalism" always involved the eating of humans as opposed to eating other giants. Many humans regarded them as ugly and stupid. Perhaps the giants' numbers increased to a point that alarmed and threatened their smaller neighbors.

A second cause comes to mind for the schism in this phase. True Giants may have lacked the capacity to advance culturally along with human beings.

Now, in ancient times, the lives of our ancestors were rough and crude in comparison to the world we see today. The cultures of Europe, for example, were not models of gentility. They sometimes rated the label of barbaric. John Grant refers to a "disgusting age" when he compares the sports of Viking and Scottish warriors centuries ago [7]. Mankind's greater tolerance for crudeness and earthiness was probably universal in ancient times. Our ancestors shared a primitive and harsh world with True Giants.

Cultural change may have exaggerated the primitive and unattractive traits of the giants as viewed by humans. The advances that humans were bound to make would have increased the differences between the two groups.

The consequence of the falling out in phase two is the deaths of many giants at the hands of human giant-killers.

The third phase of relations is the removal of True Giants to a reclusive life in remote locations. Today's giants may be regarded as primitive in contrast to their own ancestors. They might be said to have degenerated because cultural borrowing is now denied them. However, they also may be smarter than their ancestors because they have come to avoid confronting humans.

The giants we note as surviving today are all remote dwellers in mountains and areas unfrequented or avoided by humans. It is possible—even likely— that the three phases described above were acted out in regions all over the globe. Our knowledge of the European experience—familiarity, battle, and estrangement between humans and giants— may be a matter of better record keeping and the accessibility of records in European languages. Civilizations that rose and fell in many parts of the world may have had similar interactions with the giants. Today, the last phase is universal. Giants survive as remote occupants of the least hospitable terrain on the major continents.

It is time to venture behind the "ridicule curtain" that has protected True Giants from scrutiny in modern times. The giants have continued to hold out against extinction. We should consider that these giants have capabilities as well as limitations apparent in their past performance. The giants of today are the fittest of their species. They need not be the stupid giants that our ancestors defeated. Also, they almost certainly have large primate brains and the ability to apply themselves, as suggested in their ancient communications with humans.

To explain our success it is important to mention that size is not everything in brains. The organization of the brain is important. Also, it is a fact proven by our history that we humans work better together than any of the presumably big-brained man-like creatures—True Giants and others—still reported to roam the woods and mountains.

CHAPTER THREE
TRUE GIANTS IN EUROPE

OUR WORLDWIDE SURVEY of True Giants must touch all the continents except Antarctica. Giants had a colorful past in the days before history was put down on paper. They have survived in much of Europe as colorful characters in the folklore of past ages. Dietrich of Bern was one legendary hero who fought a giant in his life. The fabled Dietrich was based upon the historical person

The Fight in the Forest, pen and black ink on laid paper (1500/1503) by Hans Burgkmair I.

of Theodoric the Great (circa 454-526). He was a king of the Eastern Goths. We can see that Theodoric might have actually had to fight a giant from the mountains of Europe to gain his fame.

Many old stories merit a second look when we realize that True Giants were around to play a part in the lives of our distant ancestors. There are reasons to suspect that some of the descendants of ancient True Giants are still around today. Europeans have been wary of considering this possibility. Based on precedents, it is human nature to look first for giants in remote places like the Himalayan Mountains, and to pooh-pooh the idea that they might still be around in one's own backyard. When one's backyard includes mountain ranges, however, then one is likely to find that encounters with True Giants are being recorded even today.

The United Kingdom has never lacked stories of giants. In *The Minor Traditions of British Mythology*, Lewis Spence devotes one chapter to giants and ogres in England and a second to giants in Scotland, Wales, and Ireland [1]. According to this collection there was extensive interaction in the not-too-distant past between human beings and True Giants. The relationship reached a point, however, where the storied Knights of the Round Table were determined to kill every giant remaining in Britain [2].

Donald Mackenzie in *Teutonic Myth and Legend* touches upon what he sees as a scattered class of giants represented in Scandinavia, the middle of Europe, and particularly in Scotland:

> These are the Mountain-giants. In neglected archaic lore of Scotland they are called Fomors, but they are not the Fomors of Ireland... As river-goddesses in flight are personifications of rivers, so do these Fomors personify the hills they inhabit. Scottish mountain-giants never leave their mountains. They fight continuously one against the other, tossing boulders over wide valleys or arms of the sea [3].

Mackenzie makes two observations worthy of special note. He mentions that giants "fight continuously one against the other," behavior that is cited elsewhere in the world as limiting the numbers and success of giant beings. When Mackenzie says that Scottish giants kept to their mountains, we find a clue explaining the persistence in Scotland of giants even today. Shy mountain dwellers are the most likely to escape the extermination that was the lot of True Giants in most of Great Britain.

Scotland's vague but much discussed Big Grey Man of Ben MacDhui could be evidence of a True Giant. Some of the hairy men might be holding out on the British Isles, where they are otherwise remembered in a wealth of folklore. There was even a report of a giant footprint in 1959. Adam Young, a geologist, reported finding this print north of Ballachulish. It was in a patch of snow and measured 10 x 24 inches. He could make out toes but how many there were was not reported at the time [4].

Numerous articles have discussed Scotland's Big Grey Man of Ben MacDhui. He is held responsible for strange reports from the mountain of Ben MacDhui (4,296 ft), one of Scotland's highest peaks [5]. Climbers to that domain have reported being overwhelmed by feelings of terror. More concretely, some have heard footsteps and observed a giant figure obscured by mountain mist, hence the name of Ferla Mhor, or Big Grey Man. Here is one report of this giant:

> One witness of the Big Grey Man himself, a mountaineer, was alone at night on the mountain, and saw "a great brown creature...swaggering down the hill...it rolled slightly from side to side, taking huge measured steps. It looked as though it was covered with shortish brown hair," and he later calculated its height as between 24 and 30 feet [6].

Other Scottish peaks are associated with giants also, but the stories are of a friendlier presence. They tell of giants that appear in the form of enormous figures. There is a variety in these encounters that suggests the recent giant lore is not a single legend, but rather a record of encounters with a population made up of different individual giants.

The Gray Man of Braeriach, not to be confused with the Big Grey Man of Ben MacDhui, is the polar opposite in the nature of its presence. It is described as a friendly and helpful giant when met by human beings on the peak of Braeriach [7].

The AA *Touring Guide to Scotland* recalls for us the legends of giants that put place names on the map. One example is the community of Reidh, which got its name from "a giant who vaulted across the narrow strait from Skye to Glenelg." The *Guide* goes on to cite a local legend that speaks to our curiosity about the finding of giant bones:

> Not far from Glenelg is a place, which used to be called in Gaelic the "Field of the Big Men." About 150 years ago some gentlemen decided to dig open a great mound that stood there. The place was traditionally regarded as the burial ground of giants [8].

Skeletons of great size were found. However, a "confusion of accounts" leaves us unsure of what happened next. One of the diggers was said to have pulled from the mound an intact skull so large he was able to place it over his own head! A second account told of a medical doctor examining the remains from the mound and declaring them to be from people 8.5 to 11 feet tall. A third account advised that the excavation be refilled after an unexpected storm frightened the diggers into abandoning their project.

The giants of Scandinavia figure famously in the mythology of the region [9]. As colorful as they are, the narratives tell us little about what giants were like physically. Their similarities to humans were sufficient for them to be regarded as "giant men." For people centuries ago that was enough. Modern science, on the other hand, would examine the bones of a True Giant and conclude—in our view—that they are descendants of

Gigantopithecus.

The True Giants of Scandinavia are the Jotuns. They are the colorful characters of mythology and ancient history. The likely lands of the Jotuns in the North Atlantic are today's Iceland and Scotland.

Iceland has not been the isolated world that common histories would have us believe. It did not become part of the inhabited world only in the centuries that the Irish and the Northmen began to visit. Coins found on the island indicate that it was being visited in Roman times. It was

What reality lies behind the 550 A.D. tales of a swamp-dwelling, violent giant named Grendel in what today is southern Sweden? Drawing by Harry Trumbore.

probably the Thule of the ancient world. Knowledge of Iceland was simply forgotten for a time. There are traditions in the large wasteland region of northeastern Iceland that indicate True Giants lived there at one time [10].

Throughout Europe giants are remembered as having once been part of the population. Some may linger even today. In Poland and Czechoslovakia True Giants are mentioned as living in the High Tatras, the tallest peaks in the Carpathian chain, on the border between the two countries. The Slovaks call them Zruty (or Ozruti) and know them as "wild and gigantic beings," according to the Slavic entry in *The Mythology of All Races* [11].

At some point we should acknowledge that there are modern reports and numerous traditions of "wild men" and "wild women" in Europe and elsewhere that will not be mentioned here at all. This is because we are examining the towering giants that exceeded 10 feet in height. Some traditions of "wild men" tell of man-like beings that exceeded human beings in height, but at their greatest heights were not more than around 7 feet tall. We would identify them as Trolls and possible descendants of Neandertals. Descendants of both those distinctive and hairy near-men are known to have survived into recent times. They can be distinguished from True Giants. Their endurance in any part of Europe is a different discussion altogether.

Some traditions describing their physical characteristics are simply too vague for distinctions to easily be made. Many traditions of ogres would probably reflect the presence of True Giants if we better understood the traditions [12].

By one tradition the city of Antwerp in Belgium takes its very name from the conduct of a giant in the days of Roman might. This giant, named Druon Antigonus, exacted tolls from ships passing his castle. Those who did not pay had their hands cut off and thrown into the Scheldt River. "Hand werpen" means "to throw the hand," so the city derived its name from the legendary activities of Druon Antigonus [13].

Douai in France also has a traditional giant, this one named Gayant. The specific heights attributed to these two giants derive from effigies that remain to commemorate them, and not from any historical record of actual measurements of their sizes (Antigonus = 40 ft., Gayant = 22 ft.) [14].

In Greece giants are part of the tradition of the Callicantzari. Since Callicantzari are considered to be shape-shifters, the name is used to account for different things. Among them is "a gigantic monster whose loins are on a level with the chimney pots [15].

The island of Crete was once home to a tribe of giants who were cave-dwelling herders of animals. Modern day inhabitants of Crete still point out the places where the Triamates lived. These giants may have been the inspiration for the tribe of the Cyclops and the particular giant Polyphemos described in Homer's story of Ulysses. While attempting to re-trace the semi-mythical voyage of Ulysses, author Tim Severin came upon these stories of True

Giants on the island of Crete [16].

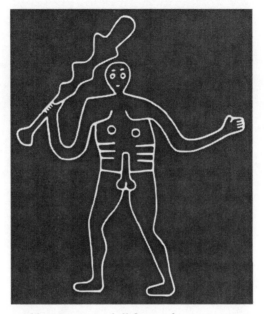

The Cerne Abbas Giant is a hill figure of a giant man on a hillside near the village of Cerne Abbas, north of Dorchester, in Dorset, England. The 180 ft (55 m) high, 167 ft (51 m) wide figure is best viewed from the opposite side of the valley. A 1996 study noted that there was once a disembodied head drawn underneath one arm. Some folk stories indicate that the Cerne Abbas Giant is an outline of the corpse of a real giant. One story says the giant came from Denmark, leading an invasion of the coast, but was beheaded by the people of Cerne Abbas while he slept on the hillside.

The ogres and wild men of European traditions indicate the past presence of True Giants. They border closely upon the recent historical period where they are scarcely mentioned. But the rare finds of giant footprints and the fear inspired in hikers upon Scottish peaks are hints that some True Giants might still survive.

Chapter Four
The Mountains of Asia
Shelter Giants

THERE ARE STORIES of giants all across Asia. Goliath of Biblical times does not stand alone. He was one of many giants known in his own time. Elsewhere in Asia the presence of high mountains seems to assure us there will be tales of True Giants to go with him. High peaks became the refuge for the Big Men as the smaller but cleverer humans overran the world.

The most famous of all giants would be Goliath of Gath. In western Asia his presence in ancient days is familiar due to his defeat at the hands of David and his slingshot. The true height of this historical Goliath has been much discussed. We are faced with arguments that make him out to be little more than 6 feet in height or towering 10 feet or better.

There were tribes of such giants known at the time. Goliath's brother Lahmi was said to use a spear the size of a tree. This is not the weapon to be wielded by a figure only 6 feet tall.

The giants of Biblical times were known as the Nephilim, the Rephaim, and the Sons of Anakim. Sir William Smith (1813-1893) ably summarizes Goliath's relations as a race of giants when defining the Anakim:

> A race of giants, the descendants of Arba...
> dwelling in the southern part of Canaan,
> and particularly at Hebron, which from their

progenitor received the name "city of Arba."
Besides the general designation Anakim, they
are variously called Sons of Anak, descendants of
Anak...and sons of Anakim. Those designations
serve to show that we must regard Anak as the
name of the race rather than that of an individual,
and this is confirmed by what is said of Arba,
their progenitor, that he "was a great man among
the Anakim"... The race appears to have been
divided into three tribes, or families, bearing the
names of Sheshal, Ahiman, and Talmal. Though
the warlike appearance of the Anakim had struck
the Israelites with terror in the time of Moses...
they were nevertheless dispossessed by Joshua,
and utterly driven from the land, except a small
remnant that found refuge in the Philistine
cities, Gaza, Gath, and Ashdod... Their chief city
Hebron became the possession of Caleb, who is
said to have driven out from it the three sons of
Anak mentioned above, that is the three families
or tribes of the Anakim... After this time they
vanish from history [1].

The dispute over Goliath's true height has arisen because
scholars have not considered the origin for the giants that we
are able to understand now. The issue of the survival of tribes
consisting of the offspring of *Gigantopithecus* has not been raised
until now. The presence of True Giants living beside humans
explains why people in ancient times had to face such formidable
opponents.

The case of Og of Bashan makes for a better example of True
Giants as remembered from the ancient world. Adrienne Mayor,
writing about "Giants in Ancient Warfare," gives this summary of
his fate:

When the Israelites finally attacked the alleged
sons of Nephilim, the formidable Og of Bashan

was one of the first giants to fall. Og had few equals in height or might, and the historian Josephus noted that he was admirably proportioned and well coordinated. Og's stronghold was an impregnable subterranean city. The book of Joshua suggests that the Israelites routed Og's forces by releasing swarms of hornets into his underground fortress. After the victory "people could get a sense of Og's strength and magnitude when they found his sleeping quarters," said Josephus. "His bed was 4 cubits wide and 9 cubits long! " (Deuteronomy, chapter three, gives the same figures.) Depending on how much leg room and headroom the giant preferred, Og could have measured at least 15 feet tall [1].

There is a famous quote often used by authors to justify the existence of giants in ancient times. The quote refers to the Nephilim and the Cainites (descendants of Cain), who were a race of giants living in an underworld kingdom. This is the oft-quoted statement to be found at Genesis 6:4, "There were giants on the earth in those days, and also afterward, when the sons of God came in to the daughters of men and they bore children to them." But this passage is an incorrect translation from the original text. What it actually noted was that there were "giants in the earth." This detail, a reference to their habitat, was a clue to how the giants had remained hidden.

Adrienne Mayor observes that such giants ceased to be a factor in ancient warfare after these defeats. A few giants were seen as continuing curiosities, however. They made appearances in Rome when the Roman Empire was in full force.

Josephus wrote that in the first century A.D., a seven-cubit-tall (about 11 ½ feet) giant from Palestine named Eleazar was sent as a diplomatic gift from King Artahanas of Persia to the Roman emperor. Other emperors displayed living giants;

Gabbaras, an Arabian giant almost nine feet tall, was a spectacle during the reign of Claudius [2].

In western Asia the Armenians knew the Torch (also Torx) to be a giant. Mardiros H. Ananikian wrote of him in *The Mythology of All Races*:

> In fact he is a kind of Armenian Polyphemos. He was said to be of the race of Pascham (?) and boasted an ugly face, a gigantic and coarse frame, a flat nose, and deep-sunk and cruel eyes. His home was sought in the west of Armenia most probably in the neighborhood of the Black Sea [3].

The Torch was remembered for "his great physical power and his daring." He crushed solid granite in his bare hands and, according to tradition, engraved images upon the stone with his fingernails. Another feat had him hurling stones after ships on the Black Sea.

The Himalayan Mountains are famous for the stories of the Yeti. The hairy mystery of the high valleys became famous also as the Abominable Snowman. Among the tales collected from the "Roof of the World" were accounts telling of True Giants. They were shunned for many years because the size of these bipedal primates seemed too large to be believed.

In the 1950s curiosity about the identity of the Abominable Snowman drew Sir Edmund Hillary into the search. The excitement had been building since the 1920s. Stories of hairy man-like creatures were known. Footprints in the snow and mud were found to indicate some such creatures were active in out-of-the-way places.

The Sherpas in Nepal know of several types of Abominable Snowmen in the mountains of Asia. When Hillary went to the "Roof of the World" to look for the Yeti, he and his collaborator, journalist Desmond Doig, noted that there were several unknown primates said to be there, still undiscovered in any formal way by scientists. Among them was the Nyalmo.

Hillary and Doig learned of the Nyalmo in north central Nepal. It was said to be "giant-sized (up to twenty feet tall), manlike, hairy, and given to shaking giant pine trees in trials of strength while other Nyalmos [sat] around and clap[ped] their hands" [4].

In the 1890s there was a reported killing of a True Giant. The incident began with some telegraph workers who disappeared at the pass of Jelap La in Sikkim. Dr. G. N. Dutt reported the results in the *Times of India* for 17 May 1953:

> Immediately afterwards the British troops stationed nearby were summoned to search for their bodies and also to account for their mysterious disappearance. A few hours' search resulted in the discovery of a Snowman, who was apparently responsible for the death of the missing persons. It was an easy target for the rifles. It was alleged to have been twelve feet tall, with shaggy hair, and toes pointed backwards [5].

An early investigator of Yeti reports named Charles Stonor also heard of this report. He dismissed the episode, apparently because of the large size of the dead creature [6].

Bernard Heuvelmans in *On the Track of Unknown Animals* gives a lengthy account of an Indian pilgrim observing giants in this part of the world. The passage appears in a 1937 book by Jean Marques-Riviere. The pilgrim had heard from chance observers of the giants that they spoke an unknown tongue. He saw them himself after joining a small expedition in Nepal that went looking for them. The group came upon footprints in mud that measured 2 feet long. At that point all but three of the expedition turned back. Eventually, following a drumming sound, the three came upon a ring of 10 giant "ape-men" 10 to 12 feet in height. One of them was beating a drum while the others moved in ritual fashion. They were described as hairy with faces that were a mixture of man and gorilla [7].

In the 1930s there were repeated instances of giant footprints reported from India. News from Calcutta on 19 July 1935 told of tracks 11 x 22 inches near Jalpaiguri. A woodcutter followed the prints. He collapsed and eventually died after catching sight of the track-maker. It was, he said, "gigantic and had a human appearance." On 20 June 1938, footprints with similar dimensions were reported in the same area by the London *Daily Herald* [8].

In 2006, the news contained the startling discovery of fossilized footprints found in southern India. They were discovered in Kerala. The impressions were of different sizes. They reached dimensions that called for the existence of True Giants to explain them. The preliminary findings were that these fossilized footprints dated back only 30,000 years.

In China giants have been included in the folklore of unusual creatures said to dwell in the region's many mountain ranges. The traditional height and tracks of these giants are double the size attributed to giants elsewhere in Asia. The description of giants in Edward Werner's *Myths and Legends of China* includes a height of 50 feet and tracks 6 feet long.

The Nyalmo are said to leave behind four-toed tracks. Drawing by Harry Trumbore.

We have been asked if we think giants 50

feet tall existed. The answer is no. We are simply passing along the information as given in Werner. We think the sizes given are simply exaggerations. They suggest knowledge of True Giants, but giants of a more modest size. Other traits included bodies "covered with long black hair" and cannibalism upon "enemies taken in battle." They populated a large area of mountains in northeastern Asia [9].

Ivan T. Sanderson in *Abominable Snowmen* reproduces a drawing of an ancient mask from the Mongolian plateau. It might be a depiction of the skull of a True Giant. Little information is given about this object. It appears to have been included in the research of Russian scientists who gathered data on unusual animals in Europe and Asia throughout the 20th century [10].

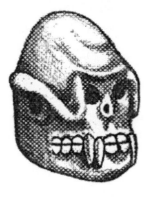 

A reconstruction of a head based upon this mask presents us with an image that is consistent with what has been described for True Giants. The face is oval in shape. The eyes are deep-set. The head comes to a peak that is an indicator of the large muscles necessary to operate the massive jaw of this ape-man. In Alaska this peak has been referred to as the "little hat" on the giants. This reconstruction is also a good match for the face of the Orang Dalam reported in Malaysia during the early 21st century.

In Central Asia an example of a footprint with consistent True Giant characteristics comes to us from the Gissar Mountains in the Pamir-Altai Range of Tajikistan. Tracks found in September of 1981 were 49 cm (19 in) long and showed four toes [11]. A photograph appeared in the *Moscow News* weekly for November 22-29, 1981. It shows a cast of a long foot with four large toes of similar size. Next to the cast is a scale in centimeters (although the width of the cast is not clearly measurable because the photo was taken at an angle). Beyond the cast is a shod human foot. The photo caption identifies this cast as a footprint found by Vadim Makarov. Myra Shackley in *Still Living?* identifies this cast as being 19¼ inches long [12].

In parts of Russia the True Giants have been noted briefly, from one end of Siberia to the other. The reports from the western end of Siberia have been rare. But the eastern portions are rich with traditions of hairy men who are feared for their appetites. There are many names for giants in the Siberian far east. The stories indicate they are enormous hairy man-like figures known to devour humans. Alexandra Bourtseva, writing in the *Technical Journal for Youth #6*, gathered together stories of giants from the Chukchee and Lamoot peoples who worked on reindeer-breeding farms. They had reports of creatures they called Mirygdy (The Shouldered One) and Kiltanya (Goggle-Eyed) [13].

One of the Lamoot people offered these descriptive characteristics for the Kiltanya: "The bridge of his nose is thin. His eyes are big. His footprint is an arm's length. His heel is thin and his toes are normal. He lives in the mountains and eats meat. He steals the shoulder blades from a deer slain by a man. He does not attack humans."

The mountains of Asia appear to provide a haven for spectacular giants that once preyed upon humans. The most recent suggestions hint that those activities are no longer taking place. When the giants stay out of sight most of the time and avoid humans, they appear to be left alone in turn by the local human populations.

CHAPTER FIVE
THE GIANTS OF SOUTHEASTERN ASIA

THE TRUE GIANTS of Asia have become a news sensation. The latest reports from Malaysia have at last called the world's attention to the existence of these primate wonders.

There have been many accounts of them in Southeast Asia. Giants there are said to reach heights of up to 20 feet and have been reported for the past 70 years. The *New Straits Times* published a photograph of a track on 12 February 1961. It was found in the province of Johor on the Malay Peninsula where giants have reportedly been seen. The foot had only four toes and measured 13 x 30 inches. [1]. Six of these monster tracks were found near Kluang in Johor. Human calls in the area brought

strange, booming responses from out of the jungle. One aborigine, Inche Yusof Kunton, reported that he had come face-to-face with a hair-covered, 10-foot-tall giant some 25 years before these tracks were found.

In this part of the world True Giants come into the news every few years and then are forgotten again until a particular report is picked up and spread around the world.

In 1971, the True Giants of Southeast Asia (sometimes referred to as Orang Mawas) were much talked about. On the Malay Peninsula, the Orang Dalam ("Man of the Interior") were reportedly seen in the provinces of Pahang and Johor. They were said to be 10 to 20 feet tall. Kurt Rolfes and Harold Stephens journeyed up the Endau River in pursuit of the creatures [2].

They have many local names. On the boundary of the states of Johor and Pahang is a large reserve of lowland tropical forest called Endau-Rompin. In this region the True Giants are also known as Serjarang Gigi. Wendy Moore has given us this account of them in a fascinating guide to Malaysia:

> The indigenous Orang Ulu who use these forests for hunting and gathering, live in Kampung Peta, just outside the park boundary on the Johor side, and at Kampung Mok, located on the Pahang side. The elders of Kampung Peta posses a rich heritage of legendary tales featuring hairy giants, fairies, spirits and dragons. For each strange rock formation there is a fairy tale to explain its shape. The uncannily quiet plateau at Padang Temambung, a shrubby grass-land habitat covered with what appears to be a criss-crossing of paths is a sacred place – the home of a Chinese fairy princess and the gateway to the Orang Ulu heaven. The landslips at Gunung Beremban have been attributed to the awakening of the earth dragon, which occurred when a careless individual brought vinegar up to the peak and aroused the sleeping dragon with the

pungent smell. Endau-Rompin is also believed to be inhabited by the Malaysian version of the "Bigfoot," known as Serjarang Gigi, "widely-separated teeth." The beast is a hairy giant, well over 3m (10 ft) tall with arms as big as a normal person's thigh [3].

In the central portion of the peninsula that is today part of Malaysia the people know the Great Spirit of the Jungle, Hantu Raya. Ronald McKie relates an interview with a former game ranger, Mat Derani, who described an encounter his uncles had with Hantu Raya. They had been attracted to him by his strong smell, which was compared to fresh spring onions:

> My uncles went to the forest to collect rotan and found Hantu Raya himself asleep beside a tree. They were very frightened. There he lay curled up like an immense sakai (aboriginal), much bigger than a man, with a lot of curly black hair and wearing only a short sarong. Although my uncles feared him, they wanted to talk to him and woke him up [4].

This was accomplished by tapping on a tree. Hantu Raya opened his eyes, which were compared to "the red of a tiger's eyes at night." Hantu Raya got up and "hopped" away down a path. This locomotion is puzzling, but, according to the narrator, Hantu Raya does not walk but "can only hop on his heels."

There is more knowledge of True Giants to the north of the Malay Peninsula. It is worth mentioning for the continuity of distribution that it suggests. In the mountains where Burma, Thailand, and Laos meet there is talk of the Kung-lu (the Mouth Man), which had terrified the people for centuries. Hassoldt Davis recorded these words on the subject:

> The Kung-lu, according to Thunderface, was a monster that resembled a gorilla, a miniature

King Kong, about twenty feet tall. It lived on
the highest mountains, where its trail of broken
trees was often seen, and descended into the
villages only when it wanted meat, human meat.
Elephants roamed hereabouts we learned, and
we were told also that no one in Kensi had been
eaten by the Kunglu for more years than the
eldest could remember [5].

Despite this anticlimax the story was interesting because it
was common all along the borders of Chinese Yunnan, French
Indo-China and Siam. Another curious thing about it was that
the Mouth Man never ate fat people, as one would expect, but
the very thinnest of them; what he liked was bones.

In Sumatra, True Giants are likely be equated to the Orang
Gadang (Big Man) about which too little has been collected to
tell us just how big it is.

The True Giants of Malaysia were in the news again in
January of 1995. There was a sighting and large footprints with
only four toes were found. The report was that wildlife officials,
police units, tribesmen, and the Malaysian army had all engaged
in a search near Tanhung Piai in Johor, but nothing was found
[6].

Frenchman Guy Piazzini went to Borneo to make an
ethnographic study of the Dyaks and the forest hunters known as
the Punan. As he was completing his book *The Children of Lilith*,
the True Giants were back in the news. He was prompted by a
brief report from Borneo in 1958 to comment on the tradition
there. He began with a reference to the long-nosed monkeys of
Borneo, one of which was shot by a hunting party [7].

In point of fact, it was only a monkey, the *pithecus
nasica*, which is, nevertheless, extremely difficult
to capture, and bears a startling and deceptive
resemblance to man. Though it is only four or five
feet tall it exactly tallies (this is the oddest thing
of all) with the description of a strange monster

44

put out a month or two back by a Djakarta press agency, which was dubbed "The Interminable Woodman."

"In the heart of Borneo," this report announced, "some river fishermen have seen a hairy, heavily bearded man, who stands nearly twenty feet in height. This is by no means the first time that such giants have been reported from the interior of Borneo."

It is, indeed, very far from being the first time; this gigantic monster figures frequently in Dyak tales and legends. We have had the Yeti, the Abominable Snowman from the Himalayas. Is it now the turn of The Interminable Woodman from Borneo to catch the public imagination?

Separating fact from fancy in this case presents certain difficulties. The opinion of the Dyaks themselves is that this is no animal, but a man, somewhat like a Punan, who lives in the jungle on a diet of wild fruit. But they also identify the creature in a strange way with some more or less mythical ancestor of their own [8].

Now we turn to the remarkable series of events that took place in this corner of the world at the start of the 21st century.

CHAPTER SIX
RUMBLE IN THE RAINFOREST:
ENTER ORANG DALAM

MALAYSIA HAS LONG been the scene of True Giant activity, but in late 2005, it all came to a head.

Right before Christmas, news dispatches filtering out of Malaysia proved to be of great interest to cryptozoologists and hominologists worldwide. Bernama, the Malaysian National News Agency, reported that there had been sightings of Bigfoot creatures in the forests of the country. The indigenous peoples of Malaysia, the Orang Asli (First Peoples), call these creatures Hantu Jarang Gigi. The advisor to the Malaysian Nature Society stated that the existence of Bigfoot in the area was certain, as there had been many witnesses. He also stated that the Malaysian Nature Society was prepared to conduct a scientific study of the creatures. The official told the media:

> Bigfoot exists. We have received reports from many people who said they had seen the creature in the forests of Tanjung Piai, Mersing, Kahang, the Endau Rompin National Park and Kota Tinggi. They [the sightings] are not a new phenomenon. In fact, I regard this as a unique feature of the Johor's treasures and we must take steps to safeguard it [1].

Sightings from the recent past were recalled. In 2001, several Forestry Department officials and campers reported seeing a Bigfoot-type creature at the Endau-Rompin National Park. Footprints were found and sightings were reported in Tanjung Piai in 2004, and encounters with the giant apelike creatures were reported at Kahang during the summer of 2005.

The first major late 2005 sighting occurred during November in Kampung Mawai, Kota Tinggi, when three workers who were building a fishpond in the village saw a Bigfoot family of two adults—who were brown, smelly, and 8-to-10 feet tall—and a child. The eyewitnesses returned to the area and saw several footprints, some large and others small, including one that was almost 18 inches long. Officials were impressed with the footprints and took the incident to be credible. Other locals came forth with information that they too had seen 10-feet-tall, black, long-armed apelike creatures while they were searching for rattan.

On 24 December 2005, new releases from Bernama stated that the Johor National Parks Corporation was preparing to undertake a scientific study to verify the claims of the Bigfoot creatures in the jungles of Johor [1]. The director of the group said that a study as proposed by the Malaysian Nature Society was needed because the sighting reports could not be taken as proof, as there is no evidence of the creatures' presence in the area. And the Johor Wildlife Department said they would install cameras in the jungle where the creatures had been seen [2].

By early January 2006, the media was in full-scale hysterics over the creatures. Reuters reported that "Bigfoot fever is gripping Malaysia, with local newspapers and the official news agency reporting sightings of a huge ape in southern rainforests." The media even wrote themselves into the story: "The Malayasian press is enjoying the story, running headlines like 'Rangers on the trail of Bigfoot' and 'Villagers' close encounter with Bigfoot.'" One Australian newspaper admitted their enthusiasm in a headline: "Going Ape Over Bigfoot." From 24 December 2005 through 4 January 2006, more than a hundred papers around the globe published articles on the sightings of unknown hairy bipedal hominoids in Malaysia. Many of these stories were also showing up on websites, in blogs, and spread in email messages worldwide [3].

Peter Loh's sketch of a drawing by a native eyewitness of the Orang Dalam seen late in 2005.

Malaysia was in the midst of a bona-fide major "monster flap," and as usual, the electronic broadcast media finally caught up to the story nearly two weeks later. Mainstream television and cable news outlets began doing reports on the Malaysian Bigfoot sightings, and the SyFy channel's "Destination Truth" would soon be there too.

Meanwhile a drawing of the Malaysian Bigfoot appeared in the *New Straits Times* at the end of the first week of January. One photograph showed a native standing on the edge of the jungle and sketching the distinctive Malaysian unknown hairy hominid, which is very different from the classic, stocky Bigfoot or Sasquatch of North America's Pacific Northwest. If the creature portrayed in this sketch is characteristic of what was seen during the 2005-2006 wave of Malaysian encounters, it reinforces the existence of a diversity of unknown hairy hominoids, and adds weight to calling this something other than Bigfoot. We feel what was being seen was a True Giant [4].

THE DATA THAT emerged in the early days of 2006 merely reinforced our suspicions that these new sightings were indeed of True Giants. On January 24th, new footprint photographs were published throughout the country [5]. The footprints appeared to be four-toed. Were they the tracks of the unknown hairy hominoid, the Orang Dalam from Malaysia? It turns out Orang

Dalam is the most frequent name associated with True Giants in this part of the world, and that may be the best local moniker for what was being seen. But, of course, modern media being what it is, with the word Bigfoot being all the rage, the news accounts kept talking about the "Bigfoot sightings."

For example, Malaysia's *The New Straits Times* reported that "biodiversity researcher" Vincent Chow had on January 23rd led a team of eight Bigfoot enthusiasts, including a professional from England, to investigate the footprints [6]. The paper said there was a strong possibility they were made by a Bigfoot creature.

The team decided these were not the tracks of elephants. But could the one animal that left footprints, but was not seen, have been a rhinoceros? Rhinos are three-toed, but they often leave prints that look like blurred four-toed tracks.

Malaysia happens to be the site of, and supports, the very rare Sumatran Rhinoceros (*Dicerorhinus sumatrensis*). The Javan Rhinoceros (*Rhinoceros sondaicus*) is apparently extinct in Malaysia and has not been seen there since 1932. The new tracks of the alleged Malaysian Bigfoot prints looked very much like Sumatran Rhinoceros footprints to us. Indeed, behind the highlighted Malaysian Bigfoot track in the photo, we noted that the rear track has the distinct appearance and structure of a rounded rhino footprint.

Regardless, the footprint photos stimulated renewed interest in the search for the giants. On January 25th the Malaysian media reported that two teams of researchers were planning to search the wilderness areas for their local Bigfoot. By now, these cryptid hominoids had been seen in the Malaysian rainforests of Tanjung Piai, Mersing, Kahang, Endau-Rompin National Park, and Kota Tinggi.

While the media waited to cover the forthcoming expeditions, various journalists penned parts of the story. Some of the best work was done by Jan McGirk in Bangkok, the UK's *Independent* reporter [4, 7]. She reported some intriguing "new" background on the original 2005 sighting:

> Malaysia has been gripped by Bigfoot fever since November [2005] when...three labourers digging a fish pond said they glimpsed a Bigfoot family of three on a river bank in Kota Tinggi reserve. They dropped their tools and fled but returned with an educated colleague to inspect and photograph the enormous footprints. A clump of brown fur, drenched with sour-smelling sweat, was also said to be recovered from the site, along with scattered fish bones [7].

This was the first time anyone had heard of any hair samples being recovered. Unfortunately, no follow-up analysis of these alleged hair samples has apparently ever occurred.

While the British press was treating the story with intelligent curiosity, the reports out of Malaysia would soon be fodder for American sarcasm. As Americans began to report on it, they did so tongue-in-cheek with increased frequency. As so often happens, television anchor people would read the reports from Malaysia and then laugh about them. One reporter on ABC News couldn't help making a reference to the most famous Bigfoot film of all time, saying that the "family of the cameraman who filmed the Bigfoot with Roger Patterson," confessed it was all a hoax. This was not true, of course, but this was typical of how the American national news outlets kept treating and demeaning the stories. The ridicule curtain had fallen on the accounts of the Johor True Giants.

OTHER THAN BIGFOOT, the Malaysian cryptid was also being called Mawas. For several years, Mawas has been a term for unknown, man-sized, hairy hominoids seen in Malaysia. Of course, what is intriguing is that Mawas in nearby Indonesia is most often related to discussions about the orangutan, (*Pongo pygmaeus*), known from ranges in the wild in Sumatra (*Pongo pygmaeus abelii*) and Borneo (*Pongo pygmaeus pygmaeus*).

It is interesting to track the usage of the word Mawas during the Malaysian flap. On January 3, for example, the *Seoul Times* reported that following a one-day expedition led by Johor National Parks director Hashim Yusof, "a database on Bigfoot or *orang mawas* sightings at various spots" was being complied.

One widely distributed wire service dispatch on January 8th, quoted director Hashim Yusoff as observing: "My personal feeling is that there is a possibility it could be what we call in Malaysia the 'mawas' ... more of a primate."

So beginning the middle of January 2006, online forums were using the term Mawas, as did the media, in two ways. One was by noting that the "tribal people call the creatures Siamang, Mawas, or Hantu Jarang Gigi." The other was by saying "Bigfoot, also known as Sasquatch, Yeti or Mawas."

The word Mawas was not the most prominent term tied to the 2005-2006 sightings and expeditions, however. Neither was the 1960s term, Orang Dalam or Interior Man, used to describe the very tall, ten-foot hairy Malaysian hominoids.

The most commonly used term to describe these creatures in broadcast media, newspaper, and online in 2006 was Hantu Jarang Gigi, which translates as Snaggle-Toothed Ghost. This moniker usually allowed the reporter to either make fun of the "toothy" name, and/or have the creature appear to live in a spooky world of phantoms and folklore. For example, in February 2006, National Public Radio's "Living On Earth" did a report entitled "Is Big Foot back? Maybe in Malaysia." NPR's report was interesting, historically insightful about the habitat, as well as significant by the mere fact that it had appeared on NPR [8]. But aside from the headline writer's mistake in making Bigfoot two words, guess what name NPR used for the creatures? The report said that "for generations" they had been called the Snaggle-Toothed Ghost, although this is hardly true. The name is rarely found in the historical literature of Malaysia.

World Magazine made the same mistake [9]: "Jungle natives have told stories about large apes for generations, though they've been called the 'Snaggle-Toothed Ghost,' not Bigfoot, Yeti, or his North American cousin, the Sasquatch."

BECAUSE OF ALL the new reports from Malaysia, an important figure in Orang Dalum history resurfaced. His name is Harold Stephens [10]. Thirty-five years earlier, Stephens had conducted the first modern expedition in search of Malaysia's Orang Dalam.

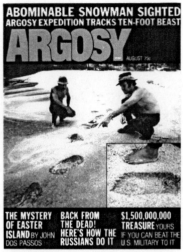

Stephens had first heard of stories of the jungle giants while on excursions into the Malaysian wilds in the late 1960s, and decided to go back with a focus on finding out more about these hairy large men. Before his expedition, Stephens had conducted

Harold Stephens shown in this photograph of the Orang Dalam tracks found on his 1971 Johor expedition.

library research and had found details on sightings in Malaysia going back three centuries.

After mounting a small expedition that included some Westerners and two local Orang Asli, Bojung and Achin, who also served as guide-porters, the team set out into the Malay jungle, up the Endau River. After reaching the 12th rapids, beyond the tributary of the Kimchin River, Stephens' party began exploring the riverbanks, looking for tracks. They found many. In 1971, Stephens found humanlike footprints that were

> enormous, 19 inches long and 10 inches wide. The creature that had made them had come down from the jungle and entered the water and here the tracks disappeared. We called the others. They came half running and half swimming across the river. Bujong came running and stopped dead. He shook his head. "Orang Dalam," he said and returned to the boat [11].

No non-native had ever found Orang Dalam tracks, and these discoveries have been the source of much discussion for nearly 40 years.

IN 2006, TWO Malaysian organizations were competing to provide information on the sightings. One was the Asian Paranormal Investigation (API), whose report ran 49 pages long, and, in essence, deconstructed the media treatment and challenged whether a Bigfoot was at all behind the sightings. The API report also used the name *serjarang gigi* for the Malaysian Bigfoot. The API report concluded that people were having encounters with Malay orangutans. Of course, cryptozoologically speaking, such a finding would still be a worthy discovery, as orangutans have not been seen on the mainland for thousands of years. But while some orangutans may have been in the mix of reports, there were undoubtedly sightings of True Giants as well.

The other organization was the Johor Wildlife Protection Society, which concluded:

> The adult creatures are between 10 and 12 feet tall while their children are 6 to 7 footers. Seventy percent of the Orang Lenggor have a human appearance but the rest resemble apes... They like to eat fish and fruits they gather in the jungles, including durian. They also have a liking for river water that contains dissolved salt and would walk for miles to get it [13].

A key here is durian, the ape's favorite fruit we already know a lot about. Was the society talking about orangutans (*Pongo pygmaeus*), a species that currently inhabits only the islands of Sumatra and Borneo? The orangutan was once found throughout Indo-China, Malaysia, and north to China. Fossil evidence discussed by anthropologist Ralph von Koenigswald, who is famous for his work on the *Gigantopithecus*, suggested that during the Pleistocene, orangutan distribution extended from Java in the south, across mainland Asia, and reached up as far north as China. Could these Malaysian Bigfoot reports be a relict population of *Pongo*?

Bernard Heuvelmans tracked orangutan-like cryptids from Assam, Burma, China, and Vietnam, through the literature, and felt (in his 1986 checklist) that there could be mainland orangutan populations in other locations than Sumatra and Borneo. Would Malaysia's Orang Lenggor turn out to be orangutans?

The Orang Lenggor reportedly has black hair on its body when young but the hair gradually turns brown as it grows older. The coloration of the orangutan (*Pongo pygmaeus*) can vary from a reddish-brown to a nearly black coat and its hair length varies considerably.

The society announced their discovery, they said, because foreigners armed with sophisticated equipment were entering the Johor jungles to track down Bigfoot without the knowledge of the state government. "We are worried these foreigners might find Bigfoot and then announce [these primates] to the world as their discovery," the society told the media [14, 15].

The society reported they were planning to field an expedition to the Orang Lenggor colony in either March or April 2006. But the society never presented any evidence of their claims.

ON 22 FEBRUARY 2006, it seemed the floodgates opened up on the hunt for the Malaysian giant apes [16, 17, 18].

Reports from a London paper indicated that a team of "paranormal investigators," including an Australian tracker, a filmmaker from Los Angeles, a reality film crew, and apparently a British reporter, was in hot pursuit of the hairy giants. This, of course, was the SyFy channel's "Destination Truth" team.

The team was lead by Josh Gates, the actor and reality show contestant born 10 August 1977, who was in the reality TV series "Beg, Borrow & Deal" (2002). The actor was also the "Butler" in the movie *Party Foul* (2003) and the still photographer for the film *The Vest* (2003). Josh Gates in Malaysia, we all would learn later, was there as the host of the soon-to-be broadcast "Destination

Josh Gates of SyFy's "Destination Truth" with members of the Singapore Paranormal Investigations group and the Seekers.

Truth" television program in the United States.

Gates was part of an eleven-member expedition from the United States: four people were from "Destination Truth," three were from the Singapore Paranormal Investigators (SPI), and four were from the Seekers. The Josh Gates's expedition found a footprint, made a plaster cast of it, and brought the cast to the United States. Measuring 23.4 by 14 inches (60 x 36 cm), the track was found in Kampung Lukut by Gates and his party after combing 30 square kilometers of the area. They found nearly a dozen similar footprints, but most had been washed away by rain and overrun by wild boars [19, 20].

Jan McGirk described the finding of the footprint:

> It resembled a gorilla footprint, only far bigger. Gates touched it, and I noticed thrashed branches directly overhead plus a dent in an adjacent log,

about two big strides uphill from the print.

I did not notice any strong smell of (Bigfoot) body odor nor did anyone find obvious hairs, fur, or droppings. We summoned the others quickly, because SPI had a police evidence kit for making a cast and the light was beginning to fade.

The casting process was filmed by the Mandt Bros productions crew. Their program's narrator, Gates, was on camera throughout, explaining what was going on. He personally lifted the cast from the soil (with care because there appeared to be an abnormally deep heel print) [21].

The footprint would play a key role in the premier broadcast of program, but that would still be months away. The cast of the footprint impression that Gates and his crew made does have a small side extension on it. Either something was stepped upon (like a twig) or it could be evidence of a large side toe. Gate's single cast may not be a completely accurate representation of the foot that made it. And remember, a cast is not a foot.

The most skeptical reaction to Gates' Malaysian foot cast came from a rival group, the API. They claimed his track cast was nothing more than a "boar-cow," as they put it. The API had also been floating other theories for what the Malaysian Bigfoot could be, including sun bears, orangutans, rhinos, and human mistakes. The tension between API and SPI was quite apparent because SPI was on the "Destination Truth" expedition [22].

Gates would not think kindly of the popular theory that his track find was nothing more than the double stepping of a Sumatran rhino or a female wild pig.

As February came to a close, Malaysian zoologist Amlir Ayat spoke up about the name for these creatures. A reporter wrote:

It would also help if a common name was used for the Bigfoot. He felt it would be a good idea to adopt the name "Orang Dalam" used by the Orang Asli, like the name Sasquatch for the American Bigfoot, which originated from the Red Indians. Mawas is used in the Southeast Asian region to refer to the orang utan (*Pongo pygmaeus*) which can grow to a height of 1.8m. Villagers in Kampung Mawai Lama in Kota Tinggi said they were familiar with mawas and were sure it was not the same as Bigfoot [23].

Local sightings continued. Kong Nam Choy, 38, who assisted the "Destination Truth" expedition by leading them into the secondary rainforest on February 20th, was later reported to have had a sighting. Kong, a construction worker, claimed that he saw the creature in March of 2006. And that month two plantation workers at Ulu Sungai Johor also claimed to have seen Bigfoot as they were going to the river for a bath.

"Myths do not leave footprints," Vincent Chow told the media at the end of March 2006.

The situation in Johor would soon grow more confusing, as the worst possible scenario would unfold. The rumble in the jungle was about to turn into a tumble.

The unfortunate outcome was that the original, genuine sightings of True Giants would be forgotten, lost in the quagmire that followed.

CHAPTER SEVEN
TUMBLE IN THE JUNGLE:
ENTER THE HOAXERS

BY MAY OF 2006, expectations were running high for scientific confirmation of the reality of Orang Dalam. But they were short lived.

It all started with a news conference that took place on the afternoon of 4 May 2006. Vincent Chow called the conference. What he said was so incredible that even our correspondent at the conference, Peter Loh, could hardly believe what he was hearing [1].

Chow, 59, presented himself with an air of confidence that could only have come from a man who knows exactly what he's talking about and is dead sure that he had solid evidence.

In a statement to the media, Chow announced that he was going to release a book the following month that would reveal everything he and his associates knew about the Malaysian Mawas from their 11-year study of the animal. He surprised everyone when he claimed to have clear close-up photographs of male, female, and juvenile animals.

Vincent Chow said he had seen the photographs and that they could not have been faked. Chow said the photos were so clear that he could see their genitalia as well as the wrinkles on their faces. They were real animals. Chow believed them to be surviving members of *Homo erectus*.

In his description of the male, Chow noted the following (as recorded verbatim by Peter Loh):

> Head: Protruding brow with thick and bushy eyebrows. Sloping forehead. Ears large, as in some monkeys. Distance between nose and lips about 2.5 times that of a man. Nose flat, somewhat like a gorilla's, nostrils very prominent. Eyes are human-like but bloodshot and protruding out of their sockets, giving them a "sinister" look. Facial hair: almost none. Body: Broad shouldered, barrel-chested, very well-built. Adults up to 8, 9 or 10 ft. Juveniles up to 5 ft. Large adults walk with a slight hunch due to height and weight. Most hair at back of head and shoulders. Hair on body only about 3 inches (he indicated with his forefinger and thumb). Juveniles have very dark hair, almost black. Adults have lighter, reddish brown hair. As they grow older, the coloration lightens [2].

When a member of the press reminded Chow that the Sultan of Johor was skeptical about the creature's existence, Chow explained that the Sultan had changed his mind after seeing Chow's photographic evidence [3].

An elderly man who had been observing these creatures for years supposedly took the reported photos. "He took them with telephoto lenses," said Chow.

Two days later, Chow replied to critics who were vocally protesting his refusal to share the photographs of these new creatures with the media. Chow said that the photos would be published in the forthcoming book, which was said to be a joint effort between reporter Sittamparam of *The New Starits Times*; Lee Hoi Chin of *Sin Chew*, a Chinese Daily; and Vincent Chow. One edition would be in English, the other in Chinese [4]. (At one point, Chow even invited Loren Coleman to write the introduction to the English version of the book.)

Over the following days and weeks, Vincent Chow and

his associate Sean Ang engaged in a war of words with critics. They defensively called for patience, saying the pictures were in possession of the "Guardians," and batted back claims from rivals that the photographs showed nothing more than escaped orangutans [5].

Two newspapers, *The New Straits Times* and *The Star*, had been in the forefront of the developing story and reporting on the "Bigfoot" sightings since December of 2005. Now a change in coverage and treatment was clearly demonstrated by *The Star* on May 26th, with their dismissive treatment of Vincent Chow. Their story paraphrased the governor of Johor State as, more or less, calling Dr. Chow unreliable [6, 7, 8].

The Star went on to say that two teams of 10 researchers were scheduled to go into the Malaysian rainforests "to gather proof of the existence of Bigfoot," according to government leader Abdul Ghani. Exactly when this was to happen, however, was undetermined.

Abdul Ghani also stated that "since talk of the Bigfoot started, we have been compiling additional information for the past three months." He noticed that cryptotourism has picked up, with 30-40 visitors per weekend going to the village where footprints had been found in recent months.

JUNE AND JULY 2006 came and went without the book being published, though all the while Chow and Ang were leaking drawings, sketches, and reports on the alleged habits of the creatures. They even launched a new website named the "Johor Hominid." But the critics and skeptics within cryptozoology grew louder and louder [9, 10, 11, 12, 13].

Finally, on 4 August 2006, Sean Ang, apparently on behalf of Vincent Chow, decided to release on their website a small portion of the eyes of an alleged Johor Hominid from one of their alleged photographs [14].

It was not much, but it was something. Of course, it only raised more questions. Why not a total and complete release of the evidence? What was this really a photograph of? Where was it taken? When? What did it show?

Then Ang released the photo on which the eye line drawing had been based. It was obvious that many details from the photograph were missing in the sketch or tracing. This further called into question the usefulness of all previous drawings or tracings they had posted. In the end, only the photos would do.

And they did. On 4 August 2006 Ang finally released the photos. First Loren Coleman posted the photos on Cryptomundo. Then Jean-Luc Drevillon, a French hominologist, discovered their source and emailed the answer to his correspondent, Lorenzo Rossi, founder of the Italian website Criptozoo. Lorenzo then emailed the source of the Johor Hominid photos to Coleman, who revealed the hoax online. Within hours, the mystery had been solved [15, 16].

The photos were in fact from a 2003 French book, *L'Odyssee de l'espece* (*A Species Odyssey*), which was made into a documentary of the same name the same year. Drevillon had recognized the alleged "Bigfoot's eyes" from "seeing the documentary and book many times." To support his claim, a full copy of the photograph, which he had scanned from the book, was posted as well.

Within hours, Sean Ang took down the Johor Hominid website, supposedly "for maintenance," but it would never appear again [17, 18].

The story of the Johor Hominid is a cautionary tale. The hoax was about the photographs, not about the sightings, not about *some footprints*, not about the reports and tracks found in the 1970s, not about the lore from the area. But in the hopes that Vincent Chow and Sean Ang had "the real deal," the quest was lost and the authentic Orang Dalam and True Giants were forgotten [19, 20, 21, 22].

The original Malaysian Orang Dalam story remains unsolved.

CHAPTER EIGHT
THE OCEANIC ADVENTURES
OF TRUE GIANTS

THE DEEP ORIGINS and spread of True Giants will have to be defined by more fossil finds. They are likely both to precede and to postdate the known fossils of *Gigantopithecus*. These tallest of primates appear to have spread around the world. There would have been little to stop such enormous creatures from expanding their range with all the capabilities they are said to possess.

More than one giant tradition credits them with using rafts to travel over oceanic waters. That use of technology would account for the storied presence of True Giants in island environments in the Indian and Pacific oceans.

They were, however, said by traditions to have not been too bright. They were inclined to fight among themselves. Also, there are traditions of them engaging in battles with other intelligent primates who were their neighbors. With these drawbacks, they would not have necessarily flourished even when they were successful in reaching many points on the globe.

The evidence of their mimicking human behavior suggests they might have benefited from the relatively recent appearance of humans and human models of survival. Putting those models into practice would have required maintaining good relations with their neighbors. That seems to have been difficult for the giants to do, so we find them today living in isolation in jungles and mountainous regions.

Traditions describe the Tano Giant of the Gold Coast, Africa, as one who kills and eats humans and wears an old cow skin. Drawing by Harry Trumbore.

The continent of Africa has many reports of large, hairy, man-like creatures, but most do not exhibit the traits that distinguish True Giants. There are exceptions, however. One is the Tano Giant which is discussed in pre-1911 accounts from the Upper Tano River area of the Gold Coast. Another tradition that hints of the existence of True Giants comes from the East African Hadendoa. And the Wa'ab of the Bedawi is a legendary creature said to live in the hills along the Red Sea in the country of Sudan. It is enormous and man-like, but little has been recorded about it [1].

The ancestors of True Giants have been around so long that they might well turn up anywhere in the world. Even the Comoros Islands in the Indian Ocean have the legend of Red Headband, a dangerous and devouring giant [2]. The Comoros are made up of four major islands; they are famous as the site of the 1938 discovery of the first coelacanth, a living fossil fish not seen for 65 million years. The Comoros are volcanic in origin and located in the Mozambique Channel, 400 kilometers east of the African continent.

Matthew Green of Reuters reported from the islands in April

of 2005 about the islanders' belief that Red Headband was a dangerous figure in their world:

> Nobody has ever seen him and lived. Quite how the ancestors ever verified the existence of "Red Headband"—the Indian Ocean's answer to Big Foot, the Yeti and the Loch Ness Monster—is thus a mystery.
>
> What is clear is that when a volcano erupted on the largest island in the Comoros archipelago on April 17, an old story gained a new twist.
>
> Since time began, an evil spirit which appears as a giant human wearing his eponymous red headband has stalked the crater at the summit of Mount Karthala, sometimes appearing as tall as a house, or even, deceptively, as a dwarf. That's only when he's viewed from far away.
>
> "When people leave the village and they don't come back, we suspect they have seen Red Headband," said Ibrahim Ali, 60, a farmer from the mountain village of Idgikgunddzi, where night fell early and rain turned black during the eruption.
>
> "Some people say they have seen him, and he looks like a giant," he said.

The tradition is regarded as a very old one in the Comoros, one that is perhaps no longer based upon the giants' presence. Green was told of seven hunters a long time ago whose skeletons were found near the summit of the volcano. Their deaths were attributed to the giant. Even though poison gas has been known to issue from the volcano and take lives, the villagers recalled the giant tradition and chose to give the giant credit.

The island nation of Australia has been the scene of True Giant encounters. There are recent sightings to support the claim that these giants inhabit the Land Down Under.

The presence of Australian aboriginal traditions of cannibal giants has long been on record. The Jogungs were regarded as gigantic and dangerous mountain dwelling men. They had clubs and used them to kill aborigines. Also there were the Koyorowen and the Yaho (not Yahoo); they were identified as:

> two tribal names for a similar cannibalistic male monster, who, after a deceitful exchange of clubs, kidnaps, kills and roasts his victim. He dwells in mountain tops and can turn his feet in every direction so his tracks cannot be traced [3].

There are modern sightings of such tall hairy men to go along with the hoary stories of old. In the outback of the Nullarbor Plain the Tjangara, or Great Hairy Man, has been seen. Footprints up to 20 inches long have been found. One fossil hunter, Steve Moncreif, reported an encounter with such a creature in August of 1972. It was wielding a club and acting aggressively. He said it was over 10 feet tall.

In 1989 in Etadunna, southern Australia there were multiple witnesses, two carloads of bush-trekkers, who told of a male creature 13 feet tall carrying a club [4].

Those who have listened to local traditions have mentioned the past presence of giants on islands in the Pacific. The subject has not been seriously examined. The record made by Captain Robert Quinton is typical when he notes that on the

Sketch of a Tjangara seen near Etadunna, Australia. Drawing by Harry Trumbore.

66

island of Nan Madol there was a tradition of little people battling the giants for control of the island in ancient times [5]. There were actual graves on the island for the Little People, but the giants were not associated with any such monuments. F. W. Christian identifies the giants as the Kona [6].

The Hawaiian Islands are reputedly the home of the Menehune, the famous Little People of a type recently found to exist in fossil form. They have been designated *Homo floresiensis*. The name applies to the intelligent but dwarfish Little People of Indonesia. Their remains have been dated as recently as 12,000 years old. The Menehune as described by native Hawaiians are the very same kind of "hobbits." The Menehune would have voyaged across the Pacific to Hawaii just as the ancestors of the Hawaiian people once did.

The Menehune are said to be surviving even today, concealed from frequent contact with humans. There are reports of Menehunes, but the native people who see them are quite protective of this knowledge of their special neighbors. So there is seldom much of a fuss made. The encounters do not often become news reports.

With that in mind, the possible survival of True Giants in Hawaii seems more palatable. The giants could have reached the islands by raft. This method of ocean travel has been cited for them in traditional accounts of the giants from North America and South America.

The island of Oahu is the source of modern reports and traditions that indicate the presence of True Giants. There is the legend of Olomana, on the windward side of Oahu. He was a warrior of great strength and height (said to be as much as 36 feet tall). Olomana controlled a large tract of Oahu from Makapuu Point to Kaaawa. Palila, a warrior from Kauai, defeated him [7].

Descendants of True Giants might still be around. People still see tall hairy men on Oahu. But these island giants do not appear to be reaching the spectacular heights attributed to their ancestors in continental settings. Instead they have been reported to be no more than 8 feet in height. This is in keeping with the tendency for species to downsize when living for a long time in an island setting. Even a modest form of True Giant presents a marked contrast to the humans who live beside them and have glimpsed them beside the roads of Oahu.

As rescued from the files of a 1973 folklore project by Nick Sucik, people have seen hairy giants in the northwest region of Oahu, near the Waianae Range where there are designated natural areas and a forest reserve. They are known as Aikanaka.

For example, Rob Carlson, who grew up at Schofield Barracks and attended the University of Hawaii, gave this account:

> One night he had gone to the river in Whitmore Village, by the Wahiawa Mountains, with his friends to catch catfish. As they were busy putting in their traps, they heard a bloodcurdling scream.
>
> "It sounded at first like a wild man screaming in the bushes right next to us. I thought at first it was a joke by one of my friends, but we were all standing there, and the screaming was in the bushes. We looked at each other and we ran like hell up the side of the embankment. We just ran, totally panicked."
>
> As they got to the top of the hill, running down a little trail, they came to a curve in the path. As they entered the curve, they all stopped dead in their tracks. An 8-foot-tall man was walking down the trail, heading right for them. He was naked except for a cloth around his waist.
>
> "I tell you, he was coming right for us! So we turned and ran back to the river." As he stumbled down the embankment, a giant woman stepped out from behind a tree. She must have been at least seven feet tall.

In 1993 Nick Sucik talked with a schoolteacher who had grown up in Wahiawa. She told him that many people in that area claimed to have seen Aikanaka lying beside the road when as they drove from Wahiawa to Waialua [8].

"Ai Kanaka" is defined in *The Hawaiian Dictionary* as a "cannibal" or "man-eater" [9]. As an accommodation to modern conditions, it is assumed the Hawaiian giants have given up this practice.

The Solomon Islands, located east of Papua New Guinea, have their own tales of True Giants. Marius Boirayon, the research director of Solomon Anthropological Expedition Trust Board Incorporated, is the individual who has spearheaded the investigations into the Solomon accounts of what he calls the Guadalcanal Giants. An Australian, Boirayon lived and worked in the Solomons as a helicopter pilot and engineer, ended up marrying a Solomon Islander, and eventually retired there. He grew to know and appreciate the culture, the folklore, and the day-to-day interaction between the natural history of the area and its people. Before long, he began hearing stories of the giants from his friends in the Solomons, so Boirayon decided to write a book, *Solomon Islands Mysteries,* that incorporates some of these stories:

> To my understanding, there are three different species or types of these Giants. The larger and more commonly seen are over 10-foot tall, but I have come across numerous islander accounts with evidence that supports that they do grow much taller than that. These Giants have very long black, brown or reddish hair, or a mixture and when they want to have a good look at you, they pull it aside from their face with one hand. Apparently, they have a double eyebrow, bludging red eyeballs, and a flat nose wide with gapped mouth facial features, and have an unmistakeable odour, which the coastal people would once use as a sign of their presence, depending on the wind. From the large hairy type, they range down in size with reducing amounts of body hair. The smaller version, although bigger than normal human beings, are like a wild man

living in the jungle and are not as hairy as the big ones. This is the way the Guadalcanal Islanders describe them....Inherently, when they see these small giant half human people, they make efforts to kill them....There are many newspaper reports, even recently, of these hairy Giants in Papua New Guinea. One of the better-known reports was a front-page item in the PNG's main paper published in Port Moresby in 2002.

The news account, Boirayon notes in his book, involved the sighting by a dozen police officers of a group of twenty 15-foot-tall hairy giants.

I must point out, that the Solomon Islanders, and I suppose the PNG and, to a lesser degree, the Vanuatu people lack understanding that the giant race living among them is a major scientific discovery to the rest of the modern world. Whether by design or not, it is appropriate that the Solomon Island's national logo is "The Place That Time Forgot" [10].

Although the reports of giants come from all over the Islands, Boirayon points out that most of the stories come from approximately 1,000 square kilometers of heavy mountainous jungle in west Guadalcanal. He has collected firsthand accounts of the sightings of the giants, who are said to be over 15 feet tall, with some 20 feet tall, and who leave footprints around construction sites. There are kidnapping stories too. *Solomon Islands Mysteries* is filled with accounts of real life encounters with these giants and the folklore that has developed from these experiences. It took a man who married into Solomon culture to crack the layer of silence about these tall hairy beings that often exists for outsiders.

In South America we find a record of the arrival of True Giants onto that continent at Ecuador. Garcilaso de la Vega,

The Museum for Völkerkunde (Ethnology) in Kiel, Germany, displays an apparently mislabeled statue, collected in 1887 on the Tami Islands in the Solomon Sea, east of Papua New Guinea—an area far outside the range of any known apes. During 2010, hominologist Markus Bühler discovered this figurine and has associated it with the giant orangutan imagery of True Giants reported from Oceania. Photo by Markus Bühler.

when recording the story of his Incan ancestors, recorded this history. In a remote time, the giants arrived on large cane rafts at Point Santa Elena in Ecuador: "These rafts were manned by males who were so tall that a normal person hardly came to their knees, although they were quite well proportioned. All of them were bearded, wore their hair hanging down on their shoulders, and had eyes as big as saucers" [11].

The giants wore animal skins or nothing at all. They quickly consumed all the food in the region. Furthermore, they caused havoc among the Indians by killing the men and abusing the women. This period of chaos only ended when the giants were killed in a violent storm. Indian tradition held that this storm was divinely directed to wipe out the giants.

There has been at least one report of gigantic footprints found in modern Peru. Some True Giants might have come to South America to stay.

The presence of traditions of True Giants on islands around the world, from the Comoros Islands off Africa to the Hawaiian Islands in the middle of the Pacific, are indications of the True Giants' ancient roots and their evident abilities to build rafts and traverse water obstacles. They even reached South America.

CHAPTER NINE
EASTERN TRUE GIANTS

AT SOME UNKNOWN point in the distant past, True Giants entered the New World from Asia, where the bones of *Gigantopithecus* have been found. One day they will be detected in North America also. However, bones of this primate are rare, even in Asia, and North America has recently (in geological terms) been scoured by glaciers, so the surviving remains may be few in the latitudes most frequently examined. Some claims of enormous bone finds have been made, though none have been properly preserved, so we will have to wait for new finds.

The American Indians have been talking about these things to ethnographers and other non-Indian contacts for years. The Indians called the giants Big Men because of their resemblance to themselves. Often the hairy wonders were known as Cannibal Giants for their appetite that included devouring Indians. But we know now that these True Giants are not over-sized men. They are the result of a long line of their own evolutionary progress, producing primates with heights exceeding 10 feet. Their large feet stamp out tracks with four prominent toes.

This historical record goes back to the 18th century. They were hard to miss. But the early records have only recently been recognized as a foundation for showing that True Giants have always been around.

The most extraordinary record in the annals of True Giants occurred in 1829 in the Okefenokee Swamp on the Florida-Georgia border. An expedition organized to investigate the find of giant footprints actually encountered a giant. The result was a disaster for both sides. Of the nine men on the expedition, only four of them returned alive. The giant also died.

The story from 1829 is unique. A contemporary record of what occurred was taken down. The recorder heard the account from the lips of one who was there, John Ostean of Ware County, Georgia. The corpse of the True Giant was observed and measured in his presence to be 13 feet tall.

While the events made the news, we cannot say that we know they created a sensation. They remained obscure in newspaper files until Richard Day of Vincennes, Indiana, noticed the story when he was searching through his hometown newspaper records on microfilm. The account was reprinted in the *Western Sun and General Advertiser* newspaper of Vincennes. That printing occurred on 6 June 1829.

The capital of Georgia was then Milledgeville, roughly in the center of the state. The town was named for a Revolutionary War hero John Milledge (1757-1818). He had also been a governor and U.S. Senator from Georgia. The town's newspaper, the *Statesman*, reported the story of an adventure of nine men in the Okefenokee Swamp.

The famous swamp, on the border of Florida and Georgia, was even bigger then than it is now. Two people had found giant footprints there. The tracks measured 9 inches wide and 18 inches long. So a party of armed men, among them John Ostean, went searching for the maker of such tracks.

Only four of them returned. They encountered a giant, 13 feet tall, a fact they established after they had killed it. Upon hearing the sounds of their guns, the giant had approached and attacked them. Five of them had their heads torn off before the effects of seven shots from their rifles caused the giant to expire. The survivors hurriedly departed in fear of revenge from any fellow giants.

The proportions of the tracks found, the stride they exhibited ("from heel to toe...a trifle over six feet"), and the spectacular height of 13 feet are all elements that have become familiar in recent decades for True Giant reports. But this event was reported

74

in 1829. Such a record is proof that True Giants have been around for some time.

There are other early records and noteworthy American Indian accounts of True Giants.

True Giants were encountered in the mountains of the Carolinas. An item from the Boston *Gazette* in July of 1793 printed this communication from Charleston, South Carolina, of May 17, 1793:

> A Gentleman on the South Fork of the Saluda river in a letter of the 23rd sends his correspondent in this city the following description on the Bald Mountains in the Western Territories. This animal is between twelve and fifteen feet high, and in shape resembling a human being, except the head, which is in equal proportion to its body and drawn in somewhat like a terrapin [in other words, no neck], its feet are like those of a negroe, and about two feet long, and hairy, which is of a dark dun colour; its eyes are exceedingly large, and open and shut up and down its face, the hair of its head is about six inches long, stands straight like a negroe's, its nose is what is called Roman. These animals are bold, and have lately attempted to kill several persons – in which attempt some of them have been shot. Their principal resort is on the Bald Mountain, where they lay in wait for travellers—but some have been seen in this part of the country. The inhabitants call it Yahoo; the Indians, however, give it the name of Chickly Cudly.

When Scott McNabb reported this find, he noted that *ke-cleah kud-leah* in the Cherokee language would mean "hairy man/thing."

True Giants are further present among the traditions of the Indians. They are associated with specific localities in the

Carolinas such as Table Rock and Jutaculla Rock.

In the southeastern U.S., the Cherokee Indians identified a specific giant as Tsulkalu. (This name has also been recorded as Jutaculla and associated with a particular rock in North Carolina that has indecipherable markings on it.) His size was such that he could only have been a True Giant. James Mooney collected the following account that put Tsulkalu into the correct context:

> James Wafford, of the western Cherokee, who was born in Georgia in 1806, says that his grandmother, who must have been born about the middle of the last century [1700s], told him that she had heard from the old people that long before her time a party of giants had come once to visit the Cherokee. They were nearly twice as tall as common men, and had their eyes set slanting in their heads, so that the Cherokee called them Tsunil' kalu', "The Slant-eyed people," because they looked like the giant hunter Tsul'kalu'... They said that these giants lived very far away in the direction in which the sun goes down. The Cherokee received them as friends, and they stayed some time, and then returned to their home in the west. The story may be a distorted historical tradition [1].

True Giants were also known in Florida. The Everglades region had its tradition of True Giants. Robert F. Greenlee recorded one in 1945 from a Seminole informant: "There is a large hammock up north of Lake Okeechobee where the tall men live. They are as tall as trees. Some of them stand up. Though they have bones like ordinary people, no living Indians have ever seen the giants" [2].

In the east, the 17th century explorer Samuel de Champlain has been ridiculed in some quarters for taking early notice of the Gougou, a giant feared by the Gaspe Indians. His account appears to be a faithful record of a fear that was probably justified:

> There is another strange thing worthy of
> narration, which many savages have assured
> me was true; this is, that near Chaleur bay,
> towards the south, lies an island where makes
> his abode a dreadful monster, which the savages
> call Gougou. They told me it had the form of
> woman, but most hideous, and of such a size
> that according to them the tops of the masts of
> our vessel would not reach his waist, so big do
> they represent him; and they say that he has often
> devoured and still devours many savages; these
> he puts, when he can catch them, into a great
> pocket, and afterwards eats them; and those who
> had escaped the danger of this ill-omened beast
> said that his pocket was so large that he could
> have put our vessel into it. This monster, which
> the savages call the Gougou, makes horrible
> noises in that island, and when they speak of him
> it is with unutterably strange terror, and many
> have assured me that they have seen him [3].

Other names later recorded for this giant are Gugwes (among the Micmac), Kookwes (among the Penobscot), and Strendu (among the Wyandot) [4].

Tales that were formerly treated strictly as evidence of Bigfoot/ Sasquatch should be re-examined in the possible context of being True Giants. Certainly, the entire body of accounts of the Stone Giants must be reassessed [5]. In Upper New York State similar beings to True Giants were known as Stone Giants. Hartley Burr Alexander in *Mythology of All Races* writes:

> The Iroquoian Stone Giants as well as their
> congeners [a member of the same kind, class,
> or group] among the Algonquians (e.g. the
> Chenoo of the Abnaki and Micmac), belong to
> a widespread group of mythic beings of which
> the Eskimo Tornit are examples. They are...

huge in stature, unacquainted with the bow, and employing stones for weapons. In awesome combats they fight one another, uprooting the tallest trees for weapons and rending the earth in fury... Commonly they are depicted as cannibals; and it may well be that this far-remembered mythic people is a reminiscence, coloured by time, of backward tribes, unacquainted with the bow, and long since destroyed by the Indians of historic times. Of course, if there be such an historical element in these myths, it is coloured and overlaid by wholly mythic conceptions of stone- armoured Titans and demiurges [6].

In Quebec an 18th century surveyor named Joseph Laurent Normandin took note of the Indian name of "Giant River." He explained: "These people have such a fear of giants that almost every day they believe that they see some of them. They affirm that an atcheme is an extra-ordinarily tall man and that he eats Indians" [7].

Here is how Charles M. Skinner described the Windigo in this part of the world:

...worst of all is the windigo, that ranges from Labrador to Moosehead Lake, preferring the least populous and thickest wooded districts. A Canadian Indian known as Sole-o'-your-foot is the only man who ever saw one and lived – for merely to look upon the windigo is doom, and to cross his track is deadly peril. There is no need to cross the track, for it is plain enough. His footprints are twenty-four inches long, and in the middle of each impress is a red spot, showing where his blood has oozed through a hole in his moccasin; for the windigo, dark and huge and shadowy as he seems, has yet a human shape and human attributes. The belief in this monster is so genuine that lumbermen have secured a monopoly of certain jobs by scaring competitors out of the neighborhood through the simple device of tramping past their camp in fur-covered snow-shoes and squeezing a drop of beef blood or paint into each footprint. There was at one time a general flight of Indian choppers from a lumber district in Canada, and nothing could persuade them to return to work; for the track of the windigo had been seen. It was found that this particular windigo was an Irishman who wanted the territory for himself and his friends; but the Indians would not be convinced. They kept away for the rest of the season. The stealthy stride of the monster makes every lumberman's blood run as cold as the Androscoggin under its ice roof, and its voice is like the moaning of the pines [8].

Have the descendants of the Gougou, the Kookwes, and the Windigo been seen in modern times? The answer is probably yes.

There was a report from Berwick, Nova Scotia, on 25 April 1969 about "a phantom scare" in the rich farming area of the Annapolis Valley. When word spread of a tall, dark form some 18 feet tall being seen there, people poured into the valley hoping to catch a glimpse for themselves. The result was bumper-to-bumper traffic. It tied up the police who had to direct traffic rather than search for the so-called "phantom."

ON NOVEMBER 1, 2010, the Major League Baseball team known as the San Francisco Giants won the World Series for the first time since the baseball team had moved West. Previously, the San Francisco Giants (1958–present) were the New York Giants (1885–1957). The name "Giants" given to 19th century baseball teams and later football teams had its origins on the East Coast in tales of True Giants from the native peoples of the area.

There is evidence for this in one exhibit of a True Giant

True Giant named Messing that the Lenape First Peoples regarded as a protector of the forests. At the Pocono Indian Museum. Photograph courtesy of Brian Rathjen, Editor of Backroads Motorcycle Tour Magazine.

on the East Coast. Bushkill, Pennsylvania's Pocono Indian Museum has an unusual display of a ˙giant wildman named Messing of the Delaware Valley. The Lenape People regarded the giant as a protector of the forest.

When the Europeans landed in America, the Lenni Lenape lived in an area called Lenapehoking, roughly the region around and between the Delaware and lower Hudson Rivers. This encompassed the modern state of New Jersey; eastern Pennsylvania around the Delaware and Lehigh valleys; the north shore of Delaware; and southeastern New York, particularly the lower Hudson Valley and Upper New York Bay. The Native Peoples of this area passed on the traditions of Giants to the new European residents of the Delaware and Hudson Valleys, whose descendants would in turn employ the name for their sporting teams [9].

True Giants were known in the middle of the continent as well to the Omaha, Kansas, and Osage Indians. These groups lived south of the Missouri River in historic times. As best we know, their ancestors lived along the Wabash and Ohio rivers. In 1885 James Owen Dorsey of the Bureau of American Ethnology recorded how these Indians remembered giants they called the Pasnuta and the Mialushka:

> A giant race, the Pa-snu-ta, once inhabited the country where the Omahas dwell. They too used to abduct people; they are called Mi-a-lu-shka by

the Kansas, and the Osages have an account of them; they had remarkable skulls, whose vertical diameter was upward of two feet. "A few years ago, when some of the Omahas were digging a grave near the house of the ex-chief Two Grizzly Bears, they unearthed the remains of about eight large people lying in a row; the skulls were about two feet long."— Frank La Fleche (of the Indian Bureau, Washington, D.C.) [10].

In Minnesota and Quebec the giant was known as the Ojibwa's Misabe and in Massachusetts as the Wampanoag's Maushope [11, 12].

An indication of True Giants can be found among the varied giant stories of the Eskimo. They had a caution against confusing the lucky find of an edible whale carcass with the sickness-causing carcass of a giant. If the carcass had a belt then it was a giant and shouldn't be eaten! The Eskimos of Greenland also thought there were enormous giants living on the edge of the island's great icecap. But few people have been willing to listen to such stories. Eskimos on the west coast of Greenland said that enormous men lived away from the coast toward the interior of Greenland.

THE MODERN REPORTS of True Giants include these sightings and footprints:

Hallsville, Texas – 1976. Zollie Owens sighted a silver-haired creature 12 feet tall. Next to it was a smaller creature with red-tinged hair, obviously a female [15].

Corinth, Mississippi – 1976. Tracks 6 inches wide and 15 inches long were found at two sites some 4 to 5 miles apart. Photographs showed four toes clearly. Some of the prints, however, failed to register a fourth toe clearly [13]. Eleven months later 15-inch-long tracks turned up to the south in Lincoln County, Mississippi [14].

Footprints found in Acorn County, Mississippi, on March 14, 1976, along Smith Bridge Road north of US 72. Credit: Lee Ann Taylor.

Fort Gordon, Georgia – 1979. Late that year a reported encounter with a hairy creature 10 to 10 ½ feet tall took place. In the area were tracks 9 inches long and 22 inches wide [16].

Poplar River, Manitoba – 1988. John Larson was out fishing when he found footprints measuring 7 x 14 inches. They had only four toes. A 14-foot stride was noted [17].

California, Pennsylvania – 1994. Tracks were found in snow in Washington County. They measured 17 x 31 inches. Only four toes were showing. From the heel of the left foot to the heel of the right foot, the distance was 6 feet [18].

The modern situation in the eastern United States seems to be one of infrequent appearances by True Giants. Those who remain probably spend their time in the mountains where the sparse population and thick vegetation help them go largely undetected. The eastern mountains from the Carolinas to Pennsylvania are the likely home of True Giants today. They will not be around in the numbers they once were when the Indians knew them well, and when they were bothering the newly arriving Europeans. But the True Giants persist in some numbers. They are seen and their footprints still appear in yards and fields from time to time.

Occasional encounters, as in Georgia in 1979, and the finding of large footprints, indicate that they sometimes abandon those mountain regions for lower country where their presence is more easily betrayed.

Also, they appear to be traveling at times from east to west and back again. Their appearances in Manitoba suggest they are moving across Canada in that latitude. And reports from Mississippi and Texas indicate such activities are going on across the southern states into the region of the Big Thicket of Texas.

The deposits of enormous footprints along a north-south route through Pennsylvania and northwestern New Jersey indicate that True Giants travel up and down the eastern states as well.

Chapter Ten
The Giants in the American West

The place of True Giants in the western areas of Canada and the United States stands out from all other reports of hairy men. The size of these upright primates and their correspondingly huge footprints do not allow them to be confused with other mystery primates once they reach adulthood.

True Giants pop up all over the western half of the continent. They appear in modern newspaper reporting, as well as in very old newspapers. They are a common theme among the Indians as recorded by ethnographers. The activities of Cannibal Giants have been a popular subject.

As people have pursued the notion of Bigfoot out west, they have come upon stories of these bigger giants that make the 8-foot Neo-Giants seem like small actors by comparison. The results have been mixed. The idea of super tall primates has not been well received by many Bigfoot trackers. Some of them have vacillated on the issue, seeming to accept them, only to later deny the idea has any merit.

The distinction between the two groups of giants has often been expressed, but not always in the same way. In 2003, one anthropologist recorded how he had heard it put to him. He was not endorsing the idea of True Giants, but only taking note of how he had been told of them by one Indian of his acquaintance:

> I did hear a western Washington State Native
> man, middle-aged, once say that the Sasquatch

one typically sees, huge as they are, are the juveniles and that they are the only ones ever sighted because, as young'uns, they are not yet shrewd enough to conceal themselves from humans 100% of the time. The adults, he said, are never seen and are unimaginably huge, twelve feet tall or something. That was a new one on me [1].

The legendary creator and guardian of the Shoshones was a mighty giant who gave them life, food, and shelter when the Earth was young. (Artists depicted them as oversized Indians.)

Here we can say that the native views of creatures can be correct in their specifics (as to differences in size) while also being incorrect in the interpretation of what those differences mean. The relationship between these creatures is not juveniles versus adults, rather they are different species that live nearby in the western forests.

An early account from Oregon tells us something of how True Giants were living in that region at the time. According to the research of Sidney Warren, the front page of the Salem, Oregon *Statesman* in 1857 contained the account of a young man who encountered a True Giant. He was carried off to the concealed home of the giants, a cave in the mountain forests of Oregon. After a short stay, he escaped to tell his story [2].

The story was told by one who had set out with an older man to look for some lost cattle. Overtaken by darkness and weariness, the two laid down to rest. About midnight the boy was awakened by a cry. He rose and walked in that direction. As Warren tells it:

> He observed an object approaching him that appeared like a man about twelve or fifteen feet high...with glaring eyes, which had the appearance of equal balls of fire. The monster drew near to the boy who was unable from fright, to move a single step, and seizing him by the arm, dragged him forcibly away towards the mountains, over logs, underbrush, swamps, rivers and land with a velocity that seemed to our hero like flying.
>
> They had traveled in this manner perhaps an hour, when the monster sunk upon the earth apparently exhausted. Our hero then became aware that this creature was indeed a wild man, whose body was completely covered with shaggy brown hair, about four inches in length; some of his teeth protruded from his mouth like tuskes, his hands were armed with formidable claws instead of fingers, but his feet, singular to relate, appeared natural, being clothed with moccasins similar to those worn by Indians.

The giant kept his grasp on the boy's arm and continued on as before. In the wilderness he finally stopped and let out a shriek that shook the forest all about them. The cry was a signal to others to open the entrance to their concealed tunnel system:

> ...immediately after which the earth opened at their feet, as if a trap door, ingeniously contrived,

87

had just been raised. Entering at once this
subterranean abode by a ladder rudely constructed
of hazel brush, they proceeded downward,
perhaps 150 or 200 feet, when they reached
the bottom of a vast cave, which was brilliantly
illumined with a peculiar phosphorescent light,
and water trickled from the sides of the cave in
minute jets.

The boy was taken into the cave and the entrance was sealed.
While in the cave, the boy observed a scene between the giant
and a young Indian woman who had been held captive for a long
time. The female was able to pass a note to the boy that advised
him to flee or he would be devoured. The boy climbed the ladder
to the sealed trap door and dug furiously to free himself. He
possessed a utilitarian Russell Barlow knife that aided his efforts.
After slipping through a hole to the surface, he encountered a
small group of miners who were prospecting on the headwaters
of the South Umpqua River. He told his story and was directed to
civilization.

At first, when the boy returned to his family, they were
doubtful, but he was known for being modest and moral so his
story was given some credit among them.

This account gives us an idea of how the True Giants have
devised ways to live concealed in the caves of the west. In this
case they were living in the 1850s in the Cascade Range near
Crater Lake in Oregon.

In 1902 a True Giant was being pursued in Idaho. Here
are the most revealing excerpts from the 29 January 1902 report
published in the Anaconda, Montana *Standard*:

Eight feet tall, covered with hair, possessing
the semblance of humanity and fierce with the
fierceness of a wild beast – that is the description
of Idaho's newest wild man, according to the
veracious *Deseret News*. An associated press
dispatch from Salt Lake City yesterday gave

the tale in full... Since the *Deseret News* is a church paper and filled with piety, the slightest insinuation that the thrilling item from Idaho is not founded on fact is entirely out of place.

When the party of skaters on the Port Neuf river saw him approach...he tried to catch them. That he failed was due to the fact that they beat him to the wagons... Even an Idaho man is not wittingly going against an eight foot monster that leaves a footprint twenty-two inches long and six inches wide...

A peculiar feature of the Idaho wild man's track is that it shows only four toes.

The huge feet with only four prominent toes are what mark the incident in Idaho as a case of a True Giant.

In the 1920s an Indian agent and schoolteacher named John W. Burns took notice of True Giants. He gathered stories of all kinds of hairy man-like creatures among the Indians of British Columbia. Only with his efforts did people begin to keep records of giant stories. Burns anglicized some Indian names for giants to create the name Sasquatch. Burns lumped all the accounts together, including True Giants, along with other mystery primates known from the same vast area of the Pacific Northwest.

Writing on the Sasquatch in 1959, Stephen Franklin noted this incident from Vancouver Island:

> Some Indian accounts are as poetic as that of Jimmy Fraser, of the Songhees [Songish] on Vancouver Island, who recalls bumping into a Sasquatch in his youth while out hunting. The Sasquatch was 18 feet tall and hurled trees at him. "His eyes glowed like the noonday sun and the hair on his body was like moss on the rocks while his voice sounded like the roar of surf from a heavy sea" [3].

That the Indians have long known about these giants is preserved in many records. Accounts of the Big Men of the Mountains are contained in the autobiographical recollections of Susan Allison published in 1976 as *A Pioneer Gentlewoman in British Columbia*, edited by Margaret Ormsby. The story of her life in the last half of the 19th century included the account of an old Indian, Ke-ke-was, who was carried off by the Big Men of the Mountains to a cave inhabited by two giants in the area of Lake Okanagan. The giants were so tall that Ke-ke-was' head reached only to their knees. They wore clothing made of goatskin, used fire, and rolled a large stone to seal their cave at night. The Indian was kept as something of a pet until he was able to flee the cave one night through a crack at the entrance [4].

Ke-ke-was was an old man when he told his grandson and his great-grandchildren of the events. The giants were known to come down from their mountain caves, lured by the abundance of fish that the Indians were catching.

When tending a fish trap, Ke-ke-was was picked up by a True Giant and carried off. As he told it:

> Soon he began to whistle. It was the same sound I had heard in my sleep and thought it was the north wind. The Big Man calmly filled the basket with fish out of my trap, then slinging it onto his shoulders, began to ascend the mountain still whistling with all his might. Once he stopped and taking me out of his breast he took a fish and tried to cram it down my throat, but seeing me choke he desisted, and putting me once more in his bosom went on his way whistling.
>
> Peeping out of the bosom of his shirt I saw we were in a huge cave. It was dark save for the red glow of some smouldering embers at the farther end. Throwing a few twigs on the embers, the Big Man blew them until with a sharp crackling sound they began to blaze, then I saw how vast a cave we were in. It was somewhat low for its size,

and from the roof hung garlic, meat, and herbs. Taking me out of his shirt, the Big Man tied me with a rope by the leg to a log that lay near the fire. There he stood looking at me, and then for the first time I had a good look at him. Thou knowest, Torouskin, that I was ever esteemed a large man, but standing by the Big Man my head was scarce level with his knees. His body was covered with garments of goatskin and was white, and he had a long bushy beard that hung down to his waist.

Later, a second giant appeared, carrying three dead does. He was as big as the first, and Ke-ke-was was able to watch them interact. They conversed in "voices like thunder." He was held captive for a long time, but one night he was able to escape the cave. He wandered for months until he found his way back to his father's camp.

After hearing this account, Ke-ke-was' grandson, Torouskin, had his own encounter. One day he heard a shrieking whistle and hid in tall grass. He then watched a True Giant walking past his location. He told his family, "Never will I doubt the wisdom and truth of the aged, for, as thou sayest, Ke-ke-was, there are many strange things in these mountains."

The Indians of Oregon had a story of giants who were chased away and had departed by raft:

There is an Oregon tradition of an underground village of gigantic Indians on Coos Bay. They bashed each other over the head with heavy bone knives without being hurt. When the smaller Indians attacked them they fled down the river and out to sea on two rafts and never came back [5].

The recollections of hunter Russell Annabel about his youth spent in Alaska (ca. 1940) include knowledge of True Giants.

He and hunter Tex Cobb heard "much about Gilyuk, the shaggy cannibal giant sometimes called The-Big-Man-With-The-Little-Hat" [6]. Here the "little hat" is referring to the point on the top of the head of True Giants. It is a sagittal crest known to occur in primates. The prominent bone structure allows the attachment of the jaw muscles that are necessary to operate the massive jawbone that *Gigantopithecus* possesses.

Their personal knowledge of Gilyuk came after a summer spent on the Nelchina Plateau, where they met a party of Indians headed by Chief Stickman.

To the Indians, Gilyuk had never been a legend. It was an animal as real as any other in the woods. The Indians showed the two hunters the sign of Gilyuk, a recently twisted birch sapling. It was four inches thick and 10 feet tall before it was twisted and began to wilt. The Indians were afraid of Gilyuk because, as Annabel wrote, "he was a shaggy giant who wore a little hat and ate men."

The following night Chief Stickman went from his camp to a nearby lake and vanished. The only trace of him was a torn garment, the one he had been wearing. The Indians quickly left the area.

Elsewhere in Alaska, True Giants have been recorded as simply "giants," as Lt. W. R. Abercrombie did in his report on an expedition to the Copper River Valley in 1884:

> Like the Yakutat and Chilkat, Kolosh, and the Ugalentsi, these people entertain some curious ideas respecting the tribes living far in the interior of the country upon whose shores they dwell, which must from their nature be purely fabulous. Thus the Kenaitze have a tradition that not over two hundred years ago the mountains northward were inhabited by a race of giants, who occasionally made raids upon their villages, and whose personal strength and size were such that the unfortunate Kenaitze who fell within the grasp of these monsters would be seized by

their feet and killed by their heads being knocked
together, after which their bodies were stored
away in the parkees of these giants" [7].

In the 1890s anthropologist Franz Boas found True Giants
identified as the Tsufa among the Indians living along the Skeena
River in British Columbia. The giants were so big they would
pick up a beaver lodge and shake out the occupants. A beaver was
only a mouthful for them [8].

Michael H. Mason toured the North Country and wrote
in *The Arctic Forests* how the Takudh Kutchin Indians feared
the Mahoni (aka the Nahoni). The name appears to have been
borrowed from the traditions of the Nakani or Bushmen, who
are described as burly and modestly sized Neandertals. The
description given by Mason for his Mahoni, however, is a dead
ringer for True Giants:

> ...the Mahoni inhabit a specified locality, the
> mountainous region round the headwaters of the
> Porcupine and Peel River, sometimes wandering
> as far west as Kandik Creek. Therefore most of
> that country is avoided by the Indians.
>
> The Mahoni are terrible wild men, with
> red eyes, and of enormous height, completely
> covered with long hair. They live without any
> fires and, whenever possible, eat human flesh.
> The Mahoni leave man-like footprints three feet
> long, and will eat a whole birch-tree, torn up by
> the roots, only throwing away the twigs [9].

Anthropologist J. J. Honigmannn noted the giants of the
Kaska Indians as "the cannibalistic Big Man (Tenatco), who used
no dwelling but sometimes dug a hole in the ground" [10].

In modern Alaska, True Giants are still a subject of
conversation. Enormous tracks have been found, such as a series
of tracks 24 inches long. They indicate the presence of True
Giants who are otherwise only glimpsed along Alaskan roadways.

In 1997 Ray Crowe reported in *The Track Record* that Andreas Trottman and Jim Repine had learned of incidents of giants in recent years near Iliamna, Alaska.

> The natives of the area call the creature the Big Man, but are reluctant to talk about it to whites. Native parents won't let their young hunters roam in the area of Tzimna Falls, 25 miles from Iliamna, because they believe that is where the home caves of the Big Man are. They don't want to have to explain to the Big Man why their children meddled in his territory... One lady did comment, Big Men "travel in pairs, sometimes two or three," says Linda Johnson, a native born and raised there. "You see, they have families just like us. They're not too smart – like people, but they know what is going on...They know."
>
> Another individual from the village of Newhalen, "Don't let the Big Men be hunted and hounded. They live here too... When the old people of the village were young, the Big Men need to come by and poke sticks at the youngsters and throw an occasional rock. They travel at night... They eat fish. I know they steal a lot of fish from the villages. I guess they might eat berries too, but I know they don't eat meat."

While the last comment is open to question, the local attitude comes through from the people living in the Iliamna vicinity. They want the True Giants to be left alone and unmolested [11].

Other tales such as the Marukarara (Uphill People) in California seem to also describe the figure of a True Giant. Charles Edson, an early pursuer of California Bigfoot stories, had heard of True Giants from an elderly trapper. The man told Edson he had seen such a giant, one over 15 feet tall. Edson simply did not believe him because it was so much taller than the Neo-Giants he had heard about more often [12].

Once in a while, the appearance of True Giants gets some widespread newspaper coverage. This happened in Alberta in 1969. A True Giant appeared within sight of the construction site for the Big Horn Dam. The dam is on the North Saskatchewan River west of Nordegg. Several witnesses mutually supported the view they had of the tall hairy beast. They also had the benefit of trees near where the giant stood— they could judge the accuracy of their estimates of its height by measuring the trees.

On August 23rd, five construction workers reported seeing a dark figure moving about on a ridge overlooking the site. The men were Floyd Hengen, Dale Boddy, Harley Peterson, Stan Peterson, and Guy L'Heureux. The figure was as tall as the spruce trees beside it. That meant it was 15 feet tall. The figure watched the men for a while. It sat down, then finally stood up and strode away, taking long steps on its thin legs. The local Indians had been seeing a group of four such creatures in the area, but they had said nothing until the excitement was raised in August at the dam site [13].

Another episode with some notoriety took place in Montana in 1977. Again it occurred in August. Three men were chased from a hill in Belt Creek Canyon by what they described as a 15-foot-tall hairy creature.

They were airmen from Malmstrom Air Force Base, but only one of them came forward and allowed his name to be published. He was Fred C. Wilson. He took and passed a polygraph test. The other two men confirmed the story to the press but remained anonymous.

The three were with two youngsters that day. They were intending to camp on a hill above Belt Creek. A thunderstorm developed at 2 a.m. on August 20th. They decided to go back to their vehicle. On their way they heard a noise. The Great Falls *Tribune* reported what Wilson said happened next:

> "I turned [on] my flashlight and saw this huge creature standing beside a tree about 25 yards away," Wilson said... "We watched it for about 10 seconds before it moved off into the trees and

then we ran for the car."

According to Wilson, the creature was walking upright across a clearing when the men reached the vehicle. One of his companions fired two shots from a shotgun to frighten the "animal" away. "The shots were not fired at the animal, but into the trees adjacent to it," Wilson said. "We were not trying to shoot it. We just wanted to keep it away from us so we could get out of there."

But the three men said the ape-like creature charged instead and got within 20 feet of the vehicle before they drove off.

Wilson gave this account of it: "It looked like a semi-truck coming at us. It took forty-foot strides. It was hideous. It had small apish-type eyes, a flattened nose, and canine-type fangs, which showed when its mouth was open. Its face was totally covered with hair. The head was oblong" [14].

This is the very giant that was identified by the American Indians in this part of North America. Mick and Ruth Gidley in *Native American Myths and Legends* identified some of them. These True Giants have been given some ominous titles such as Crusher of People and Killers of Men.

Giants and cannibals occurred in many stories. Sometimes they had only one leg or eye, or their eyes glowed with supernatural brightness. The Washoe feared a one-eyed giant who lived in the Pine Nut Mountains near the Carson Valley, Nevada. Other tribes often had legends in which the mountains were inhabited by giants: the Kalispel, Flathead, and Coeur d'Alene told of the *Natliskeligutan*, or Killers of Men; the northern Paiute told of *Numüzo'ho* or Crusher of People [15].

According to legend, the Twawhawbitts, the giant cannibal of the Paiutes, planted Indians for food.

We have heard of tunnels frequented by True Giants and caves where they dwell in the mountains mostly undetected. Such dwelling places would be a natural refuge for these enormous hairy ape-men. There is no shortage of caves associated with them in Indian lore. In California the Miwok Indians have specified the locations of large caves that were the abode of feared giants [16].

Most recently, True Giants were in the news from New Mexico in August of 2005. Three tall and hairy creatures were reported in mid August at Nenahnezad. With refreshing clarity, the people in the know in New Mexico refused to equate the local reports of hairy figures 10-to-12 feet tall with Bigfoot. They insisted the reports made sense as being the giant they knew as Yel'tso. As Ryan Hall of the Farmington *Daily Times* reported:

> According to William Tsosie of Shiprock, Bigfoot doesn't exist in Navajo culture, per se, but Yel'tso is a giant, oafish creature similar to the Anglo concept of Bigfoot. "It means 'really monstrous,'" he said. Wallace Charley of Shiprock said

according to Navaho legend, Yel'tso is a "people-eating monster." He noted the creature is also thought to assist in bringing rain to the Navaho people. He added there was no such thing as Bigfoot in Navaho legends.

Charley, Tsosie and LoRenzo Bates of Upper Fruitland all said they had heard of the alleged Bigfoot sighting in Nenahnezad [17].

The enormous footprints of the True Giants made the news in the last half of the 20th century. Tracks over 20 inches long and showing four prominent toes were reported at Pitt Lake, British Columbia in 1965; at Snoqualmie, Washington, and Cold Lake, Alberta in 1976; at Abee, Alberta, in 1977; at Yacolt, Washington, in 1980; and at Prineville, Oregon, in 1996 [18].

The vast forested and mountainous spaces of the American

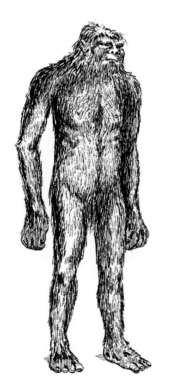

West have long been home to True Giants, according to the American Indians. Modern residents have been telling of their own encounters with hairy figures with heights of 15 feet. This record indicates that True Giants can be found anywhere from Alaska to the border of Mexico.

Two prospectors, the Welch brothers, tracked this four-toed creature near Pitt Lake, British Columbia in June 1965. They watched it for some time, used trees to estimate its height at 12 feet, and saw remarkable details, such as its auburn hair and canoe paddle-sized yellow hands. Drawing by Harry Trumbore based on the original eyewitness sketch.

CHAPTER ELEVEN
THE WORLD OF TRUE GIANTS

TRUE GIANTS RANK as the largest primates ever known to have existed. They are not, however, the top-of-the-line development among the primates. They were an experiment among primates that succeeded in the sense that such a giant can evolve and can manipulate its environment successfully. They have been around for several million years and probably longer. They are likely to have originated in the Miocene, when the ancestors of today's living hominoids were getting their start in competition for space and resources on the ground.

The mean height of True Giants is likely to be in the range of 12 to 14 feet. This was determined by Lambert Adolphe Jacques Quetelet (1796-1874). This Belgian statistician and astronomer made his judgment based upon the proposed existence of a 20-foot giant [1].

Physically they must have a unique skeletal structure. The human framework cannot grow to such heights and remain a functioning and mobile being. The bone structure of True Giants allows them to grow to heights beyond those known among humans or shown in any other living fossil.

We suspect the bones of *Gigantopithecus* will be found to have a honeycomb structure. That would allow them to develop a large size with less mass.

Such a constitution for the bones would also help explain why so few postcranial remains are left behind. None of the bones of their ancestral apemen, *Gigantopithecus*, have so far been found in scientific excavations. They might be rare, but they will not be impossible to find. Accounts of finds of arm bones, leg bones, and other body parts have been reported, but those finds were not preserved.

A range of capabilities is likely to have been found in True Giants. Some individuals were associated with humans and could have been coached to greater achievements than were likely when True Giants were left on their own. They are definitely reported to be capable of speech and to have developed their own languages. Also, their interactions with humans in legendary settings suggest they are capable of communicating in common languages with humans. True Giants will have large brains, but those brains will have been developed along an evolutionary line different from human beings.

The widespread extent of True Giants throughout the world indicates that they were once very successful. They appear to have walked and rafted their way to most parts of the globe. Eventually the competition for resources and living space pressured these primates into keeping to the high mountains and thick forests where they are reportedly seen.

Human beings were looking for ways to overcome the natural advantages that True Giants had in size and strength. One of the ways to defeat True Giants was noted among the Miwok Indians. They put sharp sticks in their paths, according to a work called *Dawn of the World*.

Another item recorded as folklore is that True Giants could be overcome with intoxicating drinks. Anthropologists have tended to see this as only a borrowed folktale, a story that was borrowed from one people and utilized by another.

A different view would be that this is a stratagem invented in different parts of the world to overcome a being of superior size. The tales that have passed along among different cultures about True Giants are often about how the big fellows were outwitted by their human captives. True Giants are portrayed as not too bright and vulnerable to tricks. The story of Ulysses and Polyphemus is an example where the giant was injured by being blinded and then fooled by tricks to allow the escape of captives.

The use of firepower in modern weapons is another way to overcome the advantages held by giants. The record for the use of sophisticated weapons begins to appear in the 18th century when European immigrants to America encountered True Giants in the Appalachian Mountains.

Modern True Giants do not have to be the very same creatures as their ancestors. We might be disappointed today if we expect to find the foolish ogres of centuries ago. For all we know, True Giants could be learning a lot from their present positions as observers of the human world.

Modern-day True Giants may not turn out to be as slow witted as their ancestors might have been. The survivors who are our contemporaries might be a lot wiser and have changed their behavior patterns from those found in old traditions.

True Giants have been under some pressure to change their ways for the past 2 million years as more and more hominids have been rising among the primate ranks to compete directly with True Giants. Because True Giants are still around, we can expect they have been responding to the competitive nature of the times to maintain their existence. That existence is different from the past when they were engaging in battles with other primates, according to the recorded folklore. They appear to have found a niche, and when people leave them alone, as we have seen the natives in Siberia and Alaska say that humans do, True Giants are living successfully beside human beings.

Gigantopithecus is a distant relative to humans in the scheme of primate evolution. Nevertheless, True Giants appear to have discovered and learned the characteristics familiar to other successful primates known as hominids. This convergence accounts for these apemen being seen by so many cultures around the world as "Big Men."

They have large hands and seem capable of making crude clothing and using simple weapons. They have engaged in some herding of animals in ancient times. They have been associated with taking up some technological pursuits such as smithing.

They are bipedal and capable of walking long distances rapidly, while taking breaks to rest when doing this. They possess binocular vision and a large brain with, it seems, the capacity for language.

They are described, at least in modern settings, as taking mates and raising families.

They seem to have a variable diet, eating plants, fish, and game animals, such beasts as caribou, deer, and beaver.

Some of the old traditions of True Giants have described them as fighting amongst themselves, though recent reports suggest there may be more cooperation among their own kind. People have reported such cooperation as in the story of Ke-ke-was from British Columbia. People in Alaska say the True Giants are now reported in groups of two and three at a time. One of the modern reports of True Giants from British Columbia is an account of two giants passing through a valley.

If we are to judge from what our study of True Giants has produced so far, greater cooperation among themselves might be their most important adaptation to the modern world that has come into being in the last few million years.

True Giants of legend were notorious as Cannibal Giants around the world. While that activity might not have disappeared altogether, it is one that makes them conspicuous and brings down a deadly response in the modern world. So that activity is likely to have faded away as the use of modern weapons spread to peoples around the world in recent centuries.

The lengthy accounts recorded by Susan Allison and a few others give us some clues about the habitations of True Giants. We know they make use of large caves and block their entrances with huge boulders and trap doors to conceal them. This has been a staple feature of giant stories throughout history. They are utilizing their strength in a way that defeats the human who wishes to gain entrance to their lairs. We have no easy way to overcome this clever use of their advantage. We cannot detect their caves easily, and even if we could, we cannot manipulate the entrances the way they do.

We have to acknowledge that they have found one important strategy that has worked well for them right up to modern times. True Giants are certainly way ahead of us in learning about their

Where can you go to see lifesize models of Gigantopithecus *in the modern world? One remarkable exhibit exists in Sauriergarten, Germany, in the older part of the dinosaur model park in Bautzen-Kleinwelka. The two other known and more conventional models are to be found at the American Museum of Natural History in New York City and the San Diego Museum of Man.*

modern human adversaries. We have collectively ignored them. We have made no efforts yet to direct upon them our modern capabilities of studying wildlife. Instead we have allowed them to remain in a limbo of animals that we ignore. They are a difficult topic to study. Collectively we have preferred to pretend they do not exist.

We expect True Giants are still evolving. Perhaps they are finding new levels of abilities that permit them to adapt and survive in a changing world.

CHAPTER TWELVE
THE FUTURE OF TRUE GIANTS

WHAT CAN WE do about True Giants?

We need to know a lot more about them before making important decisions about how to relate to them. We can start to study them in many places, for they are found all over the world. But those who set out to learn about True Giants should start listening to people who live near them.

Those would-be giant hunters should learn a lesson from what happened in Georgia in the Caucasus region of Eurasia, the homeland of George Papashvily. George immigrated to America and told the story of how he did so in a most entertaining book, *Anything Can Happen* [1]. In that introduction to his life, he mentioned a find of a cave in Georgia containing the bones of giants.

In another book he told the story of visiting again the village where he grew up. In *Home, and Home Again* he gave the full story of what happened after the startling discovery of what we can recognize as the remains of True Giants [2].

After that find, scientists happened to come to his village asking for the caves and ancient objects. But they got no help from the local people, who wished to protect these remains out of their basic respect for the dead.

As a young man George Papashvily was out one November day looking for honey with two other boys, Teddua and Bootla. They climbed higher and higher into the mountains near their village. They came to a rocky gorge where the Tergie River forms a lake. There they decided to go swimming.

The water level was low. In a cliff George spied a hole just below the surface. It proved to be a passage into a cave hollowed out of the mountain. Fissures in the ceiling allowed in a pale light: "The floor was covered with bones, bleached white, huge bones, ribs that would have a cage for a hen and all her chicks, hips wider than an ox yoke, skulls like wine jugs with eye sockets our fists could go in, bulging foreheads, huge jaws" [3].

They were familiar with the bones of common animals. They could see that these were the bones of "some kind of men." The arms had been long with five fingers on the hands. The feet had toes.

Darkness came and the boys were forced to leave the cave.

Back in their village, they approached the wise elders with their findings. The old men of the village were called Otar, Vachtang, and Miriani.

The boys were first questioned about whether they had seen swords, daggers, bowls, jewelry, or bits of cloth in the cave. They had not. They replied that the upper leg bone had been 2 feet long and the lower leg bones were 4 feet long. The elders gave their answer:

> "I think," Miriani said, "you found the Place Where the Giants Came Home to Die."
>
> "Were they men?"
>
> "Yes," Miriani said, "different from us, but men. Long, long ago they lived on earth."
>
> "Some call them Narts," Vachtang said. "Some say they were the first men. We came afterward."
>
> "What happened to them?" Bootla asked.
>
> "Slowly, one by one, they disappeared. Nobody knows where or how, for they left no graves, no bones." [4]

Vachtang said he thought his great-great-grandfather had seen one, or at least he had seen the footprint of one in the snow above Mleti. When asked if they were all dead, he replied, "I have heard a few, a very few still live at the top of Mount Kazbek."

The elders instructed the boys that the cave of the Narts had to remain a secret. The remains of the giants would be disturbed and perhaps stolen if they told about them.

The old men recalled how some very old tombs near Pasnaouri had been emptied out when found by some Russian officers 30 years earlier.

Later, some Englishmen passed through the village on their way to Mount Kazbek. They could speak some Russian, and the Russians offered to pay the locals to show them the caves. A man in the village who could speak Russian told them there was nothing for them to see there but orchards, fields, and vineyards. The people of the village perceived themselves to be protectors of the remains of the Narts.

And so the scientists returned to London, convinced that they had searched the Georgian countryside and nothing was to be found in those places. The world is not so simple a place as some people have tried to make it out to be. Even when "experts" make pronouncements from what they perceive as positions of authority and experience, they can be mistaken.

Here we see how those who have looked and found nothing can be sure of themselves. They have not only found nothing but they think they have searched as best they could.

In this case the issue was bones in a cave. Elsewhere there are many things that have been missed. Learning what others know is not so simple and cannot be achieved by offering rewards. The mysteries of the world have a place in other cultures. Those other cultures must be dealt with when we hope to get them to share their knowledge.

The record of the Narts of the Georgian mountains presents us with the kind of traces of True Giants as they turn up elsewhere in the world. There are memories of past activity, footprints are seen in the snow, sightings occur in the high mountains, and bones sometimes are found. But those important bone finds are not preserved. They confirm the stories for the local population only. Others do not pass along the best evidence for scrutiny.

We have clues about the distribution of True Giants in the past and in the present. We need to pin down their modern occurrences. They certainly survive in legends in many locales. Where they might also be in body is another matter.

The anticipated human tendency will be to think that they cannot survive in "my backyard," but maybe in someone else's. We would be better off to approach these questions with the understanding that the giants have been screened from our knowledge by their reluctance to show themselves openly and by the cultural blocks among humans who have their own wishes and feelings about the giants. These filters have kept us from becoming more aware of True Giants.

People who have lived near True Giants have made a place for them in their cultures, treating them as something special. True Giants, as in New Mexico, can be regarded as playing a positive role in the lives of the local people. In the case of the Yel'tso, the giants were associated with bringing rain. It is immaterial whether such associations are valid or not. What does matter is that the local people have such beliefs. People are influenced by their local needs.

Modern Western culture, on the other hand, has nearly universally treated True Giants as fantastic creatures lacking any credibility. We must work past that barrier and learn the facts of True Giants as best we can. We are coming late to the subject. Missed opportunities to benefit from the finds of giant bones have come and gone for centuries.

We can recall the Field of the Giants in Scotland that already produced some results according to local tradition. Bones were found that frightened the people so much they put them back into the ground. When people understand the importance of these finds, that behavior will be less likely to occur.

Their remains should be sought in caves. A study of the Miwok Indians of central California contains helpful information. *Dawn of the World* by C. Hart Merriam contains references to California caves where True Giants dwelled according to the Miwok people [5]. If such references were followed up there, and in many places around the world, we might be seeing more

Gigantopithecus remains on record.

One of the basic truths for finding bones and fossils is that one must first be seeking them out and looking in the right place before they are likely to be recognized. We need to understand that True Giants could turn up anywhere, so no place should be seen as out of bounds.

A worldwide survey of True Giant traditions should be done methodically to identify where they are known. The forests of Siberia are a location where information on them might be hard to come by, while at the same time the terrain would permit them to survive undetected for much of the time.

We need an informed overview of the worldwide status of True Giants. An organized and well-financed study of modern reports and traditional accounts, not necessarily coming from the same sources, would offer a better picture of the status of True Giants in the 21st century than has been seen so far.

There are many informed naturalists, scientists, and anthropologists around the world who know their own regional populations and the local situations. Those professionals might not want to think in terms of living True Giants, but they will have heard of them. They should be encouraged to make a record of the local knowledge of True Giants that we have described in this book. They do not have to endorse the giants. They only need to seek out and share the local views on giant primates, on ancient peoples, and on the finds of giant bones.

From other sources there will be historical records of True Giants. Those will be reports of some decades ago, as well as modern reporting of giants glimpsed and of enormous tracks seen and even photographed. These records should be assembled and shared.

The traditions, legends, and reports are all important and have the potential to contribute to a large picture of what is presently "ethno-known" (which means known to the local indigenous peoples even if unverified by science) about True Giants. This big picture is one that is made up of popular knowledge about this mystery primate. That picture will necessarily one day be subject to all manner of corrections. Those changes will be based upon further field study with dedicated search and research in pursuit of the unknown True Giants.

The topic has not been taken seriously until now. Events across the globe are changing that limited viewpoint. True Giants are being taken seriously more often today than in the recent past. The giants are under observation. Patient studies are beginning to raise the veil that has been allowed to hang over the living presence of these largest primates ever known to have existed. *Gigantopithecus* has been found. They are alive and under observation.

We now have the opportunity to learn from peoples around the world who know True Giants and have been dealing with them in the past and in the present. We must realize that gathering this knowledge requires that we respect the local conditions and the beliefs of ethnic groups who have always been aware of True Giants while much of the world has looked the other way for centuries.

ACKNOWLEDGMENTS

THE AUTHORS WISH to send out their deep appreciation to their families and friends for supporting the dreams, dedication, and creative space needed to bring the thoughts, conjectures, and research necessary to the fore to see this book come to life.

We are most grateful too for the kind permission that Bernard Heuvelmans' partner and friend Alika Lindbergh has extended to us for the use of her beautiful cover art.

Thanks also to Eric Pettifor for permission to reprint "Teeth of the Dragon."

Thanks to those who kindly have given us permission to use their illustrations and images, including Peter Loh, Harry Trumbore, Markus Felix Bühler, Brian Rathjen, Lee Ann Taylor, William Munns, Lehigh Valley Museum of Natural History, SPI/ Seekers, Joshua Gates, Neil Brandt, and Harold Stephens.

Furthermore, we could not have finished this book without our editor Patrick Huyghe's patience and assistance in completing this journey.

We wish to acknowledge these individuals and record our grand thanks, as well, to all of the hundreds of unnamed others who have helped us with the book.

Last but not least, a thank you to our readers for your courage in exploring an adventure that lives beyond fairytales and bedtime stories.

Mark A. Hall, Minnesota
Loren Coleman, Maine
October 2010

Appendix A: Telebiology

More than two decades ago, Mark Hall proposed that unknown primates be studied without harm: "With temporary captives, we should do the best we can with them and then set them free. The results will be genuine knowledge in the records we will then have, and we will have invested in the future of a new relationship with our primate relatives. This approach is part of what I have called 'Telebiology,' a means by which we can begin to study the cryptids that have been the object of cryptozoology. If we make the effort to study animals at a distance, using our brains and technology, we can succeed where others have failed in the past. If we can accept that starting to study a species with a dead animal can be difficult, then we can put that goal at the end of the process instead of making it a requirement to do anything at all." – Mark A. Hall, *The Yeti, Bigfoot & True Giants*, (1997: 110).

APPENDIX B:
WHAT SCIENTIFIC NAME
FOR TRUE GIANTS?

WHAT IS THE scientific name for Bigfoot? For True Giants? These good questions have complex answers.

"Bigfoot," of course, is the post-1958 name for the alleged unknown hairy hominids found in the Pacific Northwest of the U.S. with large, human-like footprints and an upright stance. With the Canadian form "Sasquatch," there is a longer history. The word was first coined in the 1920s (according to John Green and Ivan Sanderson) by teacher and writer J. W. Burns, who for years collected wild hairy giant stories from his Chehalis Indian friends in British Columbia. Burns created "Sasquatch" by combining several similar Native Canadians' names for these creatures.

Scientifically inclined and folkloric studies tend to use "Sasquatch" more often in recent years because it *sounds more scholarly* than "Bigfoot." Nevertheless, both are popular names and are not formal scientific names. Additionally, Mark A. Hall coined the name "True Giant" to distinguish a type of unknown primate hidden within the "Bigfoot" literature.

Various scientific names have been proposed for the animals known as Bigfoot and Sasquatch. One of the fullest discussions of this topic can be found in Grover Krantz's *Big Footprints*. What Krantz points out is simple. He notes that if he is right about his theories of what Bigfoot represents and what is evidenced in the fossil record, no new name is needed. What Krantz thinks

William Munns with his model of Gigantopithecus blacki.

and has formally written since 1986 is that, "We in fact have footprints of *Gigantopithecus blacki* here in North America." If in fact it is a different species of this genus, then Krantz would name it *Gigantopithecus canadensis*. As Krantz notes, *canadensis* "is a commonly used zoological name for species that are native to northern North America." A couple examples are *Cervus canadensis* or elk, i.e. wapiti (after the Shawnee), and *Ovis canadensis* or bighorn sheep.

Both Grover Krantz and Bernard Heuvelmans note that these are now formal assignments and proposals, and the zoological world will have to so acknowledge this if Bigfoot turns out to be a *Gigantopithecus*. If Bigfoot is in a new genus entirely, Krantz would use *Gigantanthropus*, the second name for *Gigantopithecus* that was once proposed by Franz Weidenreich in 1945, but obviously could not be and was not used. As Krantz points out, it is still available for Bigfoot. This, of course, remains to be seen, especially if an anthropologist or zoologist can make a good case that the genus discovered is so new and unrecognized that a completely new name should be given to one of these species.

Krantz further reviews a few of the possible choices if other findings prove true. *Australopithecus robustus* is to be used if these hominids are the Bigfoot; *Australopithecus canadensis* should be employed if it's a new species of the genus, *Australopithecus*. Based upon recent practice, the *Australopithecus* fossils are being routinely relabeled with their older name *Paranthropus* and some researchers now feel Bigfoot/Sasquatch are *Paranthropus*. As long ago as 1971, Gordon Strasenburgh noted that Bigfoot would be found to be related to *Paranthropus robustus*. He proposed the name *Paranthropus eldurrelli* to be specifically used for the Pacific Northwest Bigfoot.

Nevertheless, because of the standard rules of zoological nomenclature, the fact that Krantz has formally published on this topic and assigned Bigfoot/Sasquatch some possible names, if they turn out to be any of the various genus or species he covered, they have to be given one of those names. Gordon Strasenburgh's writings in the 1970s predate Krantz in the *Paranthropus* sphere, and Strasenburgh's choice would be the one if Bigfoot turns out to be a *Paranthropus sp.*

Other scientific names for unknown hairy hominoids (which include both cryptozoological hominids and anthropoids) have been formally proposed for and related to the fossil evidence. For example, the Orang Pendek has been proposed as a modern representative of *Homo erectus* by W. C. Osman Hill in 1945. The form of the yeti that is a "youth-sized ape" has been assigned the name *Dinanthropoides nivalis* by Bernard Heuvelmans in 1958. The Neanderthaloid Wildman with various local names (e.g. *almas, yeren, migo*) found throughout central Asia and allegedly evidenced by a dead body that surfaced and then disappeared in Minnesota, have formally been called *Homo pongoides* or *Homo neanderthalensis pongoides* by Heuvelmans in 1969, and Heuvelmans & Porchnev in 1974. Beginning with university lectures in 1973, and publication of the theory in 1983 and 1984, Loren Coleman formally proposed that the chimpanzee-like "skunk apes" and southern U.S. apes (which are not Bigfoot) should be assigned to the genus *Dryopithecus*.

Quite simply, what we, Mark Hall and Loren Coleman, have proposed in this book is that while Bigfoot/Sasquatch will probably be determined to be *Paranthropus*, the True Giant, if they come out of the shadows into discovery, will turn out to

117

be *Gigantopithecus*. Additionally, Hall (*Wonders*, March 1995) should get credit for bringing to our attention the finds from Greenland that anthropologists have labeled *Homo gardarensis*. If all the speculation about some of these so-called, out-of-place, more human-looking "Bigfoot types" are factual, and they do not turn out to be merely variants on the classic Sasquatch but instead are indeed *Homo*, we may have to dust off the name *Homo gardarensis*, as Mark Hall has suggested.

There has been a history of giving scientific names to these unknown hominoids. There probably will be other good suggestions tomorrow. As we know, the actual discovery and proof of these hominoids must come first, and then the names, following formal scientific protocols. The names of these creatures tomorrow, of course, are yet to be determined. The answers are not all in, however, because we are just beginning to understand what questions to ask.

Appendix C: Giant Bones

One of the most frequent questions that hominologists are asked is "Where are the bones?" In the quest for True Giants, the inquiries also include, "Where are the skeletons? Where are the skulls?" The refrain is often heard.

Are there any easy answers?

In Mark A. Hall's first two articles on True Giants, published in 1992, he proposed that the living descendants of the fossil type known as *Gigantopithecus* might still be in existence. These primates can exceed 15 feet in height and have been known as part of the cryptozoological and hominological record throughout Europe, Asia, and North America. He was careful not to assume, however, that all reports of large skeletal remains suggesting primates would have to be bones of True Giants. The bones could also belong to at least two other types of mystery primates that appear to exceed human beings in average height, but are somewhat shorter than True Giants. These other mystery primates, like True Giants, are also reported to have been seen in many provinces and states in North America.

We have been asked which finds of "giant bones" are reliable. Reliability can only be determined when the bones are preserved and the facts of their discovery and description are shared with everyone. The specter of the Piltdown hoax still hovers over the field of physical anthropology. Even professional scientists can be expected to be wary of giving an endorsement to finds beyond the current expectations in their field.

The lack of fossil finds for non-human primates is always cited as a primary reason for doubting the validity of reports of these hairy beings. But there are at least six reasons why the hominoid fossils for the various "wildmen" of North America, for example, are not already resting on scientific shelves. First, we can point out that the fossils of primates such as *Paranthropus* and *Gigantopithecus* are rare even in Asia, which is the logical continent of origin for a migration to North America. And the *Gigantopithecus* remains that do exist in Asia are meager, just teeth and partial jawbones. Fossilization is a rare event for many hominoids.

Secondly, we must point to the different cultural context in Asia that contributed to the initial discovery of *Gigantopithecus*. The giant teeth were found by Dr. B. von Koenigswald in a Chinese apothecary shop. The teeth were regarded as "dragon bones" and so something special. In Asia there is a cultural motivation for not only preserving the bones but also to put them on sale. Though there is no parallel in North America, there is some evidence to suggest that bones found were subsequently lost or discarded. Only once a fossil type is recognized, do people begin to look for more of the same. This brings us to the next reason. North Americans tend more to look for the fossils of *Homo sapiens*, rather than other primates. Gordon Strasenburgh has already made this point: "...as every sophomore anthropology major knows, those digging for fossils less than a million years old are likely to be looking for *Homo*. The site they select will relate to the habitat of *Homo*, not *Paranthropus*" [1].

Next, the populations of the other primates may not have ever been large in North America. The American Indians and the Eskimos describe knowledge of them. They appear in the post-contact records of the new Americans. But when any one of the fossil types first arrived from Asia will remain guesswork until we preserve in context some of their fossils. We simply don't know the population sizes today or in the past, 12,000 to 500,000 years ago, and the globe-girdling environments have changed drastically over geological time. The inhospitable environments of today were simply not a barrier in the past.

Our fifth reason for scarce fossils in the New World is the very dynamic nature of the planet Earth. North America has been recently scoured by glaciers. Much prehistory has been ground away under the force of ice [2]. So the best place to look for fossils of moderate age is Cuba and the Caribbean, areas that were temperate zones when a glacial ice cap rested upon Hudson Bay.

The final, and perhaps most significant, reason for the absence today of fossil specimens is that finds of "giant bones" appear to have slipped through our fingers.

Giant Bones

The record of giant bones from the New World can be grouped into three categories: (1) Bones recently found and lost; (2) Bones found early in this century and in the 19th century; (3) Bones found centuries ago.

The lost bone most familiar to Sasquatch followers is likely to be one cited by John Green in *On the Track of the Sasquatch* [3]. He learned from a woman in British Columbia how she and her husband had found some bones 20 years earlier (thus in the late 1950s). They were trapping near the Toba River at the time. It was a remote spot, meaning they had to carry out all they had. Over her husband's objections, she only brought out the jawbone; it was large enough to fit over her face. She kept it around the house and showed it to people for the next 10 years until the house burned down.

Stephen Franklin, in a mid-20th-century summary of the British Columbia Sasquatch, penned this paragraph:

> The trail of the Sasquatch is littered with accounts of discoveries of giant bones. Some were reported shipped by a Lillooet coroner to the provincial archives — and lost in transit; others were ordered tossed into the turbulent waters of the Fraser Canyon by a C.P.R. [Canadian Pacific Railway] section foreman; others are said to be lying in deep caves along Turtle Valley, east of Kamloops. None has reached competent hands [4].

Dana and Ginger Lamb reported an extraordinary skeleton

from Mexico. In the 1940s they were traveling down the western coast of Mexico. Their overland route paralleled a trip they had previously made by water. On the Rio de Baluarte in the state of Sinaloa they were shown to an Indian mound that had recently been disturbed by a flood. Artifacts were partially exposed. One of their guides, Jack Barker, observed that the next flood would probably carry away the entire mound. They began to excavate and came upon a large olla, which they described as twice the size of any oil jar from Ali Baba and the Forty Thieves. They hesitated to open the jar, but Barker advised them that his requests for museums to look at things he had found were routinely ignored. They made an opening by lifting out cracked pieces of the jar. Inside they could see two skeletons, one small and one that looked gigantic. The finger bones were twice the length of Dana Lamb's own. The size of the shin bones looked to be also twice as long. He thought the skeleton could be that of a person 8 feet tall. They replaced the pottery pieces. Later they wrote a letter to the museum in Mexico City to notify them of the site [5]. In all probability, as Jack Barker predicted, the mound was simply washed away by the river that same year.

You might think that the mystery of finds of giant bones would be helped if only one bone could get into the hands of a professional anthropologist. Surely, we would suppose that in today's world, any genuinely new bone would be recognized, preserved, studied, and then celebrated as a revelation. Unfortunately, recent history argues against such expectations. You are more likely to see extreme professional caution, shuffling around of the specimen, neglect, subsequent denial of any knowledge of it, no record-keeping (if it were important there would be a record), and the eventual loss of the specimen. We write of these things because they have already happened. You will find all of these responses detailed in the history of the Minaret Skull [6].

In brief, this partial skull was found in August 1965 by a medical doctor, Robert Denton, while backpacking in the Sierra Mountains of California. The find appeared to him to be human but was unusual for its size, shape, and "markings." He passed it along to a pathologist in Ventura County. That gentleman passed it on to two archaeologists at the University of California, Los Angeles (UCLA), Herman Bleibtreu and Jack Prost. Journalist Alan Berry learned of it in 1973. With the help of the pathologist, Gerald Ridge, Berry tried to locate the skull and learn its fate.

Blethtreu and Prost had both moved on to other jobs. Prost was contacted and flatly denied any knowledge of such a skull. Bleibtreu in 1973 had already denied any knowledge when queried by a colleague. Berry contacted him and read a letter Ridge had written to Bleibtreu. At that point he remembered the skull and said it was "certainly unusual." He was sure the skull was in the collections at UCLA. Berry had already sought the skull from museum officials, but they claimed to have no record of it and pointed to Prost or Bleibtreu as having it. The skull had essentially disappeared.

Matt Moneymaker pursued this report and learned the likely disposition of the find. There is a museum warehouse in Chatsworth, California, that receives submissions from all over the world. The finds are put into crates and stored until someone calls for them. It is likely that the unusual bone was put into a crate and marked with a number. Without knowing the number, no one can request the proper crate [7]. This recalls the classic final scene from the film *Raiders of the Lost Ark*. In the film, a valuable artifact is simply stored away and lost in a vast warehouse of crates. That seems to have happened to the Minaret calvenum.

We can naturally expect professionals to be cautious about committing themselves to new discoveries. We have no argument with that. But the history of the Minaret Calverium suggests we should expect the worst from the specialists in anthropology. They may be too scared to do anything constructive.

Other finds are lost as well. Mark A. Hall cited two instances from Minnesota in a 2002 article in *Wonders* [8].

Some bones found in the Boundary Waters Canoe Area in 1968 were described by an anthropologist as possibly representing an isolated population of primitive men in the New World. But by the time the discovery was made public, the bones had been misplaced. The University of Minnesota said they shipped them to the Smithsonian, and the Smithsonian said they had no record of them. This is no rumor, but another example of slipshod professional behavior.

The second instance was a find that ended up at the James Ford Bell Museum in Minneapolis. Samuel Eddy had retrieved the bones from a bog in Minnesota. They remained unlabeled in his private collection of odd bones, until they too were dispersed and vanished. This is the dark side of professional science, where people make mistakes and then refuse to talk about them. The participants eventually die, and then people know little about the cases, do not have to deal with them, and can contend, unjustifiably, that the finds do not and never did exist.

The second category of alleged giant bones are the many claims for bones found earlier, especially in the nineteenth century. As one example, in 1974 Borden Burleson of Cashion, Arizona, told Robert Thomas of the *Arizona Republic* about such bones [9]. His family had lived in Mexico for three generations. According to his family's history his grandfather, Alexander Burleson, had dug up "human skulls" comparable in size to basketballs. This occurred in the western Sierra Mache region of Mexico in 1898. Borden Burleson had spent some time in the area trying to find evidence of giants. He was quoted as saying, "Both the Yaquis and the Tarahuinara Indians have legends of a race of giant Indians that were said to be there when their forebears first came to the area."

Around 1951 the magazine *Doubt* published a collection of old claims to bone finds from Minnesota [10]. Duluth resident Jack Clayton gathered the stories from newspapers and such sources as *The Aborigines of Minnesota*. Numerous writers selling "true mysteries" have borrowed from Clayton's article in the past fifty years, often without giving him credit. Similar collections of old news items can be found in other parts of the country, such as a page from *The North Jersey Highlander* for Spring 1973 [11]. The editor, W. Mead Stapler, reports finding three references to giant skeletons and teeth in New York.

There are three newspaper reports from the 1930s that have — like Clayton's items — been a mainstay of "true mystery" writers. This is because the stories appeared in the *New York Times*, which provided brief accounts of giant finds in Sonora, Mexico (2 Dec 1930), in Nicaragua (14 Feb 1936), and in Florida (9 June 1936).

We do not assume that all claims are correct in their identity. The truth can only come from a close examination of each claim. A good example of what can be done occurred when a Canadian zoologist pursued an old account of fossil whale bones. They had been found in 1906 and were mentioned in a book of local history. Richard Harrington spent a year tracking them down and found them in a barn loft [12]. The value of old accounts is that they might lead to a valid find when pursued.

Bones once thought to be giants may well turn out to be some other kind of animal. Willy Ley's *Exotic Zoology* contains accounts of European finds ("giants' bones") that turned out to be the bones of extinct elephants [13]. Francis Buckland tells how a "giant in an Irish bog" turned out to be the headless remains of an Irish elk [14]. But it would be as unjustified to assume that all the claims are mistaken identity as it would be to assume that they all are genuine giant primates.

One of the worst cases of scientific bungling is to be found in the history of *Homo gardarensis*. The first scientist to scrutinize this 1926 find of bones correctly saw that they were something remarkable. But following his death another opinion was favored, and the bones remain neglected to this day. Prof. F. C. C. Hansen of Copenhagen believed the bones belong to a Troll, but when Hansen died no one bothered to study the find. Instead, the safe and comfortable way to catalog the bones was to declare the individual to be a case of acromegaly. That was the view adopted by Sir Arthur Keith, and no one else dared to contradict this lofty dismissal. But Mark Hall has pointed out since 1995 how that view is inadequate and wrong. The finding of the remains of a Troll where the Norse colonists had recorded Trolls to reside makes sense. There is also an extensive record of such creatures known to the Eskimos. All that has been ignored in favor of pigeonholing the find as nothing more than diseased bones.

Even when we have bones in hand to support our view, that is not enough in the face of scientific lethargy.

The oldest accounts of giant bones may be the least useful.

We need only note that they are not absent. In South America finds of giant bones are claimed as evidence of giants in Ecuador [15] and elsewhere on the same continent, in what is now the southernmost region of Bolivia [16].

We appear to have squandered our opportunities in the New World to identify the remains of giants, and we don't know how often this has happened elsewhere.

Having reviewed some of the history of giant bones, let us not repeat the mistakes of the past. Primatologists look for and find *Gigantopithecus* only in Asia. In the New World only *Homo* is sought and that is what is found. In the Americas giant bones found by anyone else are not going to be given a fair hearing. You can, if you like, call this the seventh reason why no non-human and recent primate fossils have been verified in the New World.

Professional anthropologists are paralyzed by fear and ignorance. The fear was explained to Mark Hall in 1968 by a teaching assistant, Robert Lynch. Hall was taking an introductory course in physical anthropology when he and Lynch had a discussion about "Bigfoot" and the Roger Patterson film then in the news. Lynch advised Hall that scientists were not going to look into "Bigfoot." He cited the Piltdown hoax, which had embarrassed many British anthropologists. He told of how Americans in the field were fearful of getting drawn into another hoax of any kind. He also advised that new ideas were simply unwelcome in the real world of science. While he was personally supportive of Hall's curiosity, he provided examples of people who were discouraged from pursuing new ideas. One example that Hall still remembers was of a dentistry student in that very university. When he had proposed a new procedure for a particular dental problem, he was told that there was already a procedure in place and that he should forget his idea.

The paramount reason there are no fossils for "giants" of any stripe is that no one is looking for them. Should we ever stop collectively stumbling through the New World landscape, either ignoring or losing the unusual bones there, we should be ready for some surprises.

Just as the full impact of the discovery of the Hobbits, *Homo floresiensis*, in Indonesia has not fully been realized, even with bones in hand, so too will the future have to wait for the bones of giants to become a formal part of the picture of hominoid history.

APPENDIX D: GIANT SKULLS BY
IVAN SANDERSON

THERE ARE SOME things I can readily accept; there are others, however, over which I boggle, or from which I retreat precipitately. I have been in full retreat from this one for nearly eight years, but I am afraid that, on the grounds of common honesty, I must now throw all caution to the proverbial winds—i.e., the storm of criticism—and give it to you straight. This whole business at first sounds so balmy as to constitute an absurdity ... but....

In 1961 I wrote a book entitled *Abominable Snowmen: Legend Come to Life*, which covered just about all that has been said on the matter of relic, primitive, fully-furred hominids still existing in various parts of the world from Scandinavia and the Caucasus, throughout the mountainous areas of Asia, to Siberia, then into Alaska, and south all the way down the western sides of North, Central, and South America to Tierra del Fuego, as well as all across northern Canada, and in various places in the tropics. As a result, all manner of interesting little ditties came to me in the mail. Most of them simply confirmed what I had said other people had said about these things, or added similar items, but from other areas. Among them, however, was a remarkable letter from a lady in Idaho—remarkable not only for the information it contained but also for its extraordinary cogency and the manifest demonstration that the writer was a person of not only higher education, but of rather exceptional erudition. Among other things (after denigrating herself almost to the point of near intellectual extinction) she announced that she had just published a book on the hybridization of Irises. You don't, or cannot, do that unless you know something of genetics. I would like to publish this letter in full, but space does not allow, so I must paraphrase.

The facts came from one of her sons (she was a great-grandmother) who had been an engineer in the U.S. Army during World War II. This man relates the following: Having volunteered in 1940 for active duty, he was sent to join an engineering unit that built the Alcan Highway to Alaska. When this was completed, he was sent, with this unit, the 1081st Company, Maintenance Engineers, to the island of Kodiak for a rest period, and was then shipped with his unit to a tiny island named Shemya that lies half a mile east of Atu (and which is separated from it only by a half-mile shallow channel) that is the last of the Aleutians going towards Asia. The Japanese were still on Atu and the purpose of landing on Shemya was to turn the island into an airstrip, it being flat and low, except for a small rise at the eastern end. Enemy resistance had been expected here but, on landing, only one dead Japanese soldier was found. However, there were neat signs all around the island stating that it, and anything found on it, was the property of (of all things) the Smithsonian Institution! When these signs were erected was not known to this engineering outfit—whether they were pre-war and left by the Japanese, erected by the enemy, or by some military unit that had got there before them. This business is odd to say the least; but wait.

According to my correspondent, her son stated that when the bulldozers arrived, they started leveling the whole island of small bumps and finally tackled the slight elevation at the east end. Curiously, this was said to have been composed of many layers of "muck," silt, and soil, with underlying sedimentary rock, while the lower land and the beaches were composed of a mixture of sedimentary and non-sedimentary rocks and boulders. As this eastern bump was scooped off, bones of all kinds began to come to light, first, those of whales, seals, walrus and such, but later and lower, those of extinct animals like mammoths. Finally, at a depth of about six feet, what appeared to be a graveyard of human remains was uncovered. These were wholly of crania (not whole skulls) and the long bones of the legs. Associated with them were numerous doll-like artifacts carved out of mammoth and walrus ivory, but "fossilized"–after they had been carved. There were also chipped flint instruments (no flint on the island) and other bone and stone implements of both very small and a rather large size.

The crania of the human skulls, which are categorically stated to be of modern human conformation with full foreheads (not sloping, ape-like ones with big brow-ridges) measured from 22" to 24" from base to crown. What is more, every one of them is said to have been neatly trepanned!

Now, the average person's skull measures only about 8" from front to back, and the cranium, i.e., the upper bit containing the brain box, stands only about 6" high–and we measure an average 5 feet 6 inches tall. Of course, there can be small people with very big heads, and there can be enormous people with small heads. I once crossed the Pacific on a Japanese liner with a Texan who was then alleged to be the tallest man in the world, at 9'2" in height. Of course, he wore a 10-gallon hat, so that the size of his head could not be accurately ascertained. He also wore cowboy boots, but he was indeed impressive, and had to enter the main saloon on all fours. He was also a charmer, especially to the Japanese flight attendants who did not reach his waist, and the Captain who did–just. However, the proportions of the body to the head in the case of a cranium that stands nearly two feet tall are something quite else again. Such an enormity is this that we resorted to some practical investigation by blowing up the outline of a modern-type human skull, enlarged to the measurements given, but on the conservative side of 22" high. The proportions and size of the body needed to support this item—of a humanoid form—would be something that stood about twenty feet.

Now, a large male giraffe may stand almost twenty feet, and the extinct Baluchitherium, the bulkiest land animal we know—the dinosaurs not excluded—which was related to the rhinoceroses, also stood as tall and had a gigantic and massive body. However, both these animals are supported on four legs. A humanoid of this type would presumably stand on only two, and, while its bulk would be less than that of either of these other animals, gravity would still exert an enormous pull (or push) down upon these two legs. How could the creature get about, even with enormous leg muscles? There is a limit to the tonnage in air that bone can support on the surface of this earth (in air that is) and, although bone is an amazingly strong material, it has to become progressively more massive to support weights above a certain point; and there would seem to be a point beyond which it simply cannot go, lest it become so massive that it literally bogs down the whole animal. But in water ...

The record whale ever measured was a female Blue at 113½ feet; and by the new method of estimating total weight at 1½ tons per foot of length, this comes out at about 170 tons. This enormity probably could, like its confreres, leap clean out of the sea, but if stranded it would die of suffocation in short order, since its sheer weight would crush the rib-cage and lungs. Buoyed up by water, however, the gravitational pull on its mass was completely nullified. The same goes for all other animals that live in water — fifty-foot squids, one-ton jellyfish, six-foot lobsters, and so on.

Now, if we must accept this report of human-like beings with crania 22" high, and thus needing a massive body some twenty feet tall to support them, what would be the most rational solution of their problem? It would be for them to live in or spend most of their time in water.

There are two aspects to this mad exercise. First, a highly esteemed scientist of the utmost probity, Professor Alister Hardy of Oxford University, England, made so bold as to publish a technical paper in 1960 on the possibility that (modern) man went through a semi-aquatic stage by gaining his food by diving for shellfish off shallow coasts [1]. A note in this paper suggested that he (man) had retained head-hair to protect his scalp from the sun. This notion at first sounds almost as balmy as our present exercise, but this scientist was neither ridiculed nor read out of court. In fact, he was taken seriously by many of his colleagues (this is something that has never ceased to amaze me). The other aspect of the suggestion that, if twenty-foot men ever did exist, they must have lived in the sea, and this leads us into other channels. We will start with the word "kelp".

This word is defined by the dictionary as: "Large kinds of seaweed; calcined ashes of same, used for the manufacture of carbonate of soda, iodine, etc., formerly used in making soap and glass." "Kelpie," on the other hand, is a Gaelic word now incorporated into the English language, but meaning originally a "Water-Spirit, usually in the form of a horse, reputed to delight in the drowning of travelers, etc." (Note the somewhat ominous "etc.".) From the former designation we derive our North American name for the vast beds of seaweed that grow in comparatively shallow waters all along our west coast from the farthest western Aleutian Islands, via Alaska, to southern California, which local citizens call simply the "kelp beds." These are very remarkable in many respects, not the least being that some of their vast fronds that float at the surface of the sea are anchored to the bottom by stalks that may be nearly half a mile long. In these kelp beds there exists a large and varied fauna; these range from specialized invertebrates that cling or buzz about in its floating fronds to the Gray Whale, several seals and sea lions, and the remarkable Sea-Otter. Most of these animals are predaceous or carnivorous, and they find a wealth of food in the kelp beds.

A race of twenty-foot-tall humans could not obtain however, even in this environment, enough animal food to maintain themselves. Lacking the cutting-teeth of the seals, or the scoop-mouths of whales, which ingest tons of small food, or grasping appendages, they just would not have been able to gain a living. If, on the other band, they were vegetarians and fed principally on the kelp itself, they could indeed have thrived and multiplied, and grown to such monumental proportions. Then there is another thing.

There is now undeniable evidence that, whatever the cause— the earth's crust is shifting; the axis wobbling, or the whole earth is going through successive cold and warm phases—the far northern latitudes around the Bering Sea, Alaska, and eastern Siberia once, and until comparatively recently, enjoyed a warm temperate climate. On Wrangel and other islands north of eastern Siberia there have been found, in addition to endless bones of mammoths and other mammals, whole flowering and fruit-bearing trees, notably of the order of the plums, up to forty-feet in length, buried shallowly in the muck around their coasts. This whole area in fact seems to have been habitable for a long time by animals evolved in warm temperate climates. Could humanoids, hominids, or even humans have developed the practice—as mooted by Professor Hardy—of gaining their living by diving in moderately warm coastal seas? Could they have continued to do so, while the general climate deteriorated, by leaning most heavily on kelp for food? The idea is admittedly most highly improbable but can we honestly say that it is impossible?

So, one has to turn to the third and last aspect of this whole preposterous business. This is to say, to the "historical."

When the lady in Idaho wrote me those four pages of most sensible material, I immediately replied, asking for further information. She replied, saying that her son positively refused to write on this matter and for several reasons: notably that an Englishman (whose name is very well known in the literary field) had annoyed him to the point of complete withdrawal by writing demanding letters of a patronizing nature that infuriated him. However, my correspondent wrote to her son on my behalf and obtained the name and number of the military outfit in which he worked in Alaska, the Yukon, on Kodiak, and on the island of Shemya. I then began a process of checking, working through the General Services Administration, National Personnel Records Center, St. Louis, Missouri.

From this most estimable and competent organization, I obtained the names of four officers of this 1081st Company, including that of their senior Intelligence Officer. I began writing letters. I received most gratifying replies from two of these gentlemen, one whom confirmed that he was with the outfit on Shemya but stating that he had not heard of any anthropological or archaeological discoveries there. The other letter, from a gentleman now resident in New Jersey, stated: "I recall that as we were building a road around the south east end of Shemya Island, the bulldozers did uncover some human bones, ivory carvings, etc. There was considerable excitement over this. . . . I recall that this area was put under the control of the Base Commander and all of the findings were to be handled by this base unit." The other two retired officers to whom I wrote did not reply. Later, however,

I traced down the Senior Intelligence Officer of this unit, but my letter to him was returned, stamped "Moved—No Forwarding Address."

To go back, though, I find that I should report some much less pleasant implications. First, there is this curious business of the island being clearly marked "off-limits" as being the perquisite of the Smithsonian. I do not quite understand this. But then comes a much less pleasant conundrum. It is alleged by my primary informants that the men aboard the island made a sort of hobby of collecting the artifacts found with the bones, but that they were told to turn them all in, under penalty. However, one man who had been a museum preparator, knowing something of their value and possible significance, made a small collection that he hoped to take back to the mainland. This was discovered, and the man was immediately arrested and held incommunicado. Later, when a civilian crew of engineers came to relieve the enlisted outfit, this man was allegedly shipped back to the States "in irons," as the saying goes, and was dispatched to (the military) Leavenworth.

Then come a number of flat statements from various sources; to wit, that a number of these skulls, or bits of them, plus other bones, some of the "dolls," and other artifacts, were collected, crated, and dispatched to the Smithsonian. I have no evidence that this was (or is) so, apart from these written statements. However, now thoroughly irked by all this, I made formal application to the Smithsonian for some clarification of all this—either a written denial of it, or some information as to just what happened to any material of this nature that was shipped to them from the Island of Shemya, circa 1945-46. I have never received a reply.

Either this whole story (and I would emphasize that it is just that, rather than a "report!", as of now) is pure hogwash, or it is true. If the former, how come such very sensible-sounding persons have written as they have; and how is it that there is confirmation, up to a point, from ex-military personnel who were at the spot when this happened? If it is true, then where the hell are the finds? Why have they not been examined, published upon, and otherwise made public? As my original informant said in one of her letters: "Perhaps you are right in saying that these people just cannot face rewriting all their textbooks."

But the really unpleasant thing to me is being asked to accept anything so utterly bizarre as twenty-foot, semi-aquatic, marine, "modern" humans. Isn't this pushing things a bit too far: or is it? I have to await expressions from the Smithsonian if there are any, which I am afraid I have to say that I rather doubt at this juncture.

Meantime, we reconstructed the outline of the alleged Shemya crania. This was done simply by extrapolation or "blowing-up" the outline of an average modern human cranium. I then asked an old friend of mine, the anthropologist Professor George A. Agogino, currently head of the Paleo-Indian Institute, Eastern New Mexico University, Portales, New Mexico to come take a look at the photo. He is one of the very few professional scientists who we felt would not burst out laughing and then refuse even to listen to the story. In this we were correct.

George Agogino said nothing when he first saw this monstrosity. He regarded it for a very long time; then asked if he could go into another room and read the file undisturbed. This he did; when he rejoined us, the first thing he asked was what the shaded outline was within the last molar tooth. I had not pointed this out to him earlier as I wanted him to have the story straight and without this entirely extraneous interjection.

This sort of "inner tooth" in the drawing is an outline-tracing (actual size) taken from a photograph of the first tooth of an extinct creature, named *Gigantopithecus*, originally found in a Chinese apothecary store in Hong-Kong by one Dr. G. H. R. von Koenigswald [2] in 1935. (Since then quite a number of teeth and some bones of this giant anthropoid have been found in caves along with other deposits in southern China.) From the conformation of the teeth and bones it is now generally thought that this creature was a giant Pongid or ape. Reconstructions of it have been published, notably one in the *Illustrated London News* [3]. The animal was assessed at eight to twelve feet tall by the British, ten to fifteen feet tall by the Chinese.

Reducing the outline of the cranium to fit this inner outline, we then found that we had a skull of such enormous size as to be quite beyond belief. Since this tooth exists, there can be no question about its size per se. (The alternative is that the creature it grew in had a jaw out of all proportion to the rest of its head, like a Pithecanthropine, an Australopithecine, or some lower type, or like the crazy Olduvai skull turned up by Leakey.) However, its owner cannot be a vast ape because this tooth is typically hominid, and even "human" shaped!

We then tried the whole thing over again with a molar tooth of another extinct hominid named *Meganthropus* (thought to be a large form of Pithecanthropine) and therefore to have had a very small brain box in proportion to its jaws; but still the skull, patterned on a blowup of Pithecanthropus itself, was so enormous it would have required indeed a twelve-foot body to support it. Thus, we found ourselves going around in a circle of speculation. George Agogino was gracious enough to hear us out and comment on each of these efforts without either lapsing into ribaldry or bypassing the "logic" of the exercise. But he did admit to being greatly puzzled by one thing. This was that nobody seemed previously to have "speculated" upon the implications of the sizes of the teeth of *Gigantopithecus* and *Meganthropus*.

Just what kind and size of skull did they grow in? Further, did the hominids develop huge vegetarian forms that needed these enormous teeth for crushing rough fibers? Finally, could any such forms have had truly "modern" human-type skulls? There are no answers to these questions, and there will not be until and unless we get substantial parts of said skulls, and of the limb and other bones of the bodies that supported them.

Appendix E: The Toonijuk
by Ivan Sanderson

The Eskimos of today maintain a large body of tradition about a race of very primitive people with revolting habits who occupied their territories prior to their own arrival. This tradition spreads all the way from Alaska to Greenland and throughout the Canadian Arctic Islands. These creatures are said to have been very tall, fully haired, dim-witted and retiring; but to have fought savagely among themselves, been carnivorous, and to have gone naked, though they built circular encampments of very large stones with whale-rib and skin roofs. The Eskimos say they had primitive stone and bone implements. They are referred to today on Baffin Island and north to Greenland as "Toonijuk" but are called by many different though similar names to the west.

This tradition has been reported upon by many, including Rasmussen and, most notably, by Katharine Scherman in her *Spring on an Arctic Island*. Rasmussen has even stated that some of these creatures existed in Greenland within the current century but were driven up into some "inaccessible valleys" by Eskimos. This, as Scherman has pointed out, seems hardly credible, since the interior of that country immediately behind the narrow coastal strip is an ice cap. However, there are still large areas of Greenland not fully explored despite massive air-travel over much of its periphery. Also, the extreme north, around the Cape Maurice Jesup area, is not glaciated and is extremely hard of access over land, and even from the sea, due to its fjord-like topography.

These Toonijuk are said by the Eskimos to have been of giant size and to have had some exceptional and, to them as well as to us, disgusting habits. They are said to have preferred rotten meat and, it is alleged, their females tucked meat under their clothing (?) to promote decomposition by their body-warmth. Further, since they did not know how to cure skins, they are said to have wetted them and then worn these raw to dry them; and then to have used them for bedding. Perhaps the most peculiar custom ascribed to the Toonijuk, as reported by Scherman, is that young men were sewn up in fresh seal skins containing "worms" (maggots?) which, by sucking their blood, reduced their weight and so made them fleet, lightweight hunters. These maggots are believed by the Eskimos to have been fostered in the rotting carcasses of birds and one such—an auk—was said by Rasmussen to have been discovered in Greenland in his time and to have been declared by the local Eskimos to have been left there by a party of Toonijuk who, they said, had only just fled back into these "inaccessible valleys" of the interior.

While regarded as being utterly primitive, the Toonijuk are said to have lived in underground houses (though without sleeping platforms) and to have had pottery—or at least "cooking pots"—and some weapons. In Greenland, the Eskimos say that they went naked but that their bodies were covered with feather-like fur; in more westerly areas, they are said to have used skin clothing. Everybody agrees that they were very good hunters; could call game by voice or gesture; and were so strong that they could back [sic] an adult Bearded Seal. In addition to these details, Scherman records—from information obtained from the Eskimos of north Baffinland, as transcribed by P. J. Murdoch, an agent of the Hudson's Bay Company, who speaks fluent local Eskimo—that the Toonijuk were not dangerous to the Eskimo but, to the contrary, were very timid and cowardly, and were particularly afraid of dogs, which they apparently did not understand. All agree that they fought a great deal among themselves, but some Eskimos assert that their own ancestors hunted the Toonijuk down individually and so eventually exterminated them. Yet, Greenlanders insist that even today some linger on in their country but that they are excessively wary—in fact, more so than animals.

Scherman further notes that: "Until 1902 an extremely primitive tribe of Thule people lived on Southampton Island, and some of their customs were those (alleged to be) of the Toonijuk."(The Thule along with groups named the Dorset Islanders and the Sarquaq, constitute known previous inhabitants of the Canadian Islands and the far north.) Scherman (1955) herself visited what was then stated by the Eskimos of Baffinland to be a Toonijuk settlement on Bylot Island, and gives a clear description of it.

In a small isolated valley her party was shown a series of circular mounds. These proved to be composed of very large stories half buried in the permafrost. Each circle was dug out and had obviously once been roofed; they were entered by what had been a three-foot high tunnel; were paved with large flat stones; and had stone benches at the back. Around the walls were very old rotten bones of the Greenland Right Whale. The party was greatly impressed by the ability of the original builders to have dug so deeply into the permafrost with only crude stone and bone implements; and, even more so, by their having transported these enormous stones, which were not of local origin, even if they had had the use of dogs and sleds. Their Eskimo companions told them that the Toonijuk could lift rocks that no Eskimo could handle; that their houses were roofed with whale ribs; and that two whale jawbones were placed on either side of the entrance tunnel. However, this site, as Scherman remarks, showed abundant signs of having been occupied by Eskimos for long and frequent periods since its original construction.

It is most significant to note that the description of these round-houses coincides very closely with the Neolithic "Round-Houses" of the Shetlands, Orkneys, and the Hebrides off the coast of Scotland, which also were circular, sunk about three feet, surrounded by stone walls that rose some three feet above the ground, and had domed roofs made of a "wheel" of large whale ribs over which skins, peat-sod, or other insulating material was placed. The Eskimo still make stone igloos with ingeniously constructed roofs of overlapping stone slabs and which also have tunnel entrances—but they are of nothing like the size described; nor do the stones of which they are built in any way approach the size of those used in the structures said to have been built by the Toonijuk.

But of even more interest is the description of a nearby cairn of very large stones, which had partly collapsed. The interior of this is said to have been hollow, and in it lay a number of large human bones. One of the party leaned in and extracted what is said to have been a female pelvis; but, as there were no professional anthropologists in the party, they very properly replaced this and closed up the cairn to the best of their ability. Scherman quite rightly makes a strong plea for this site to be visited by competent experts and thoroughly examined before such potentially priceless relics finally disintegrate; and she ends by asking the pertinent question "Aside from the Toonijuk, if they ever existed, who else could have been here?"

Her only other thought is that they could have been Norsemen, whose sturdy build and stature, greater than that of the Eskimo, coupled with their propensity for feuding, might have given rise to legends that in time became transferred from one alien race to another; and she ends with the extremely significant remark that there were traditions and apparently detailed knowledge of White Men among the Eskimos long before recorded history.

Appendix F:
The Teeth of the Dragon
by Eric Pettifor

(This article represents the conservative anthropological view of Gigantopithecus. *Though a brief mention is made of Sasquatch and the Yeti, nothing is said of True Giants, a subject far beyond the outer frontiers of scientific respectability.)*

In ancient Greek mythology a hero named Jason yoked two fire breathing bulls and plowed a field. Into the furrows he sowed dragons' teeth from which sprang men (Hamilton, 1942).

The Chinese have for centuries sold dragons' teeth and bones to be ground up as a medicinal. These bones are actually ancient fossils. In 1935 G.H.R. Von Koenigswald discovered a fossil tooth in an apothecary shop in Hong Kong (von Koenigswald, 1952). Since then 3 jawbones and over a thousand teeth have been recovered, not only in apothecary shops but in situ as well (Ciochon, Olsen, & James, 1990). They are the remains of an extinct ape, *Gigantopithecus blacki*. There are sites where *Gigantopithecus blacki* remains occur along with *Homo erectus*, such as at Tham Khuyen in Viet Nam, and in the Hubei and Sichuan provinces of China (Ciochon et al., 1990). At Tham Khuyen the remains of a potential competitor for bamboo, a proposed major food source of *Gigantopithecus blacki*, were found as well: the giant panda, now extinct in Viet Nam (Ciochon et al.,

1990). *Gigantopithecus* teeth from Wuming, China, have been dated to the middle Pleistocene, around 400,000 B.P., by faunal association. *Homo erectus* was in Asia by that time and may have played a role in the extinction of *Gigantopithecus* (Ciochon et al., 1990).

Appearance

According to Ciochon et al. (1990), *Gigantopithecus blacki* was 10 feet tall and weighed 1,200 pounds. This is speculative, since it is with some uncertainty that one reconstructs such a massive creature from a few jawbones and teeth, however many. The way they arrived at this picture was first to estimate the size of the head from the jaw, and then to use a head/body ratio of 1:6.5 in order to determine the body size. For comparison they cite a head/body ratio of 1:8 for the *Australopithecus afarensis* specimen known as "Lucy." The more conservative ratio for *Gigantopithecus* was arrived at out of consideration of the massive jaw as an adaptation to the mastication of fibrous plant matter (probably bamboo). *Gigantopithecus* was probably proportionally a markedly big-jawed creature. For the head shape they based their assumptions on the orangutan, since evolutionarily they place *Gigantopithecus* on the same line as the orangutan, finding a common ancestor for them both in *Sivapithecus*. However, the orangutan could not serve as a model for the body, since it is unlikely that a 1,200-pound ape would be as arboreal. Therefore they chose the largest primates known, the gorilla and the extinct giant baboon *Theropithecus oswaldi*, as their models for the body. They gave *Gigantopithecus* an intermembral index 108 (gorilla at 120 + *Theropithecus* at 95 divide by 2 = 108 rounded up—very scientific!) (Ciochon et al., 1990).

Since Ciochon (et al, 1990) with the aid of Bill Munns (Hollywood monster maker/dinosaur reflesher) were interested as well in building a very impressive life-size model, we would be wise to consider the dimensions with some caution, and note that they represent the biggest *Gigantopithecus* that could be built rationalized from the actual remains, and that it is a male. Females may have been half the size of the males, since the teeth fall markedly into two distinct size groupings (Ciochon et al., 1990), as I will discuss later in terms of sexual dimorphism and what inferences have been drawn.

Elwyn L. Simons and Peter C. Ettel (1970) paint a somewhat different picture. They trace *Gigantopithecus* back to a dryopithicine origin and their corresponding reconstruction is essentially a giant gorilla, 9 feet tall, weighing 600 pounds. It is not nearly as attractive as the giant orangutan/gorilla cross created by Ciochon et al. and Bill Munns (1990).

Sexual Dimorphism

Simons and Ettel (1970) do go into greater detail regarding the mandibles, however, and speculate that the size differential between two of them (Mandibles I and III) reflects sexual dimorphism. The way that the teeth fall into two distinct categories was discovered by Charles Oxnard, an Australian anatomist, when he analyzed 735 *Gigantopithecus* teeth. All teeth from the first incisors through the third molars occurred in both groups in equal numbers (Oxnard, 1987, cited in Ciochon et al., 1990). Furthermore, the size differential is greater than that occurring in any living primate including both gorillas and orangutans. Ciochon (et al., 1990) note that in living species this usually indicates competition between males for multiple females, but go on to note Oxnard's argument that the equal numbers of males and females suggests general promiscuity free from competition. "The resultant increased proportion of females pregnant at any one time under such a system (perhaps almost all of them), together with harsh environmental conditions, including fierce predator pressure, could combine to produce small inter- or intra-sexual selection, but strong sex-role differences and therefore strong sexual dimorphism." (Oxnard, 1987, cited in Ciochon et al, 1990). This sounds good, but does not address the fact that even in species with marked sexual dimorphism and sexual competition, males and females will be born in more or less equal numbers and can reasonably be expected to leave behind equal numbers of teeth. It seems that this is an instance where complex social behaviour is difficult to determine solely from physical remains, especially remains as regrettably incomplete as those of *Gigantopithecus*. If there are analogies to be made with living primates exhibiting marked sexual dimorphism, equal numbers of surviving male and female teeth cannot be a factor in the analysis.

Geographical Distribution

Geographical distribution is likewise sketchy, since the majority of remains are from one site, Liucheng Cave in Liuzhou, China, though there have been other finds in Viet Nam and in China, so that we may define southeast Asia as the range of *Gigantopithecus blacki*. A separate species of *Gigantopithecus*, *Gigantipithecus giganteus*, was found in northern India, but this specimen predates *Gigantopithecus blacki* by about five million years, and there is some controversy as to the exact nature of its relationship. Simons and Ettel (1970) place it as directly ancestral to *Gigantopithecus blacki*, while David W. Frayer (1972) argues that it is ancestral to the Australopithicines, only to be refuted by Robert S. Corrucini (1973) on the basis of multivariate analysis and so on. Physical remains for this species are even rarer than for *Gigantopithecus blacki* and the opportunity for speculation and statistical gamesmanship is correspondingly greater.

Locomotion

Ciochon et al., (1990) speculate that given its size *Gigantopithecus blacki* was a ground dwelling ape, probably a knuckle walker, though it could just as easily been a fist walker, the exact nature of its locomotion is impossible to ascertain from mandibles. Given its mass it could not have been a gibbon-like brachiator.

Diet

When considering diet, the teeth can provide us with stronger clues via analysis of opal phytoliths.

> An alternative technique [to analysis of wear patterns and other conventional methods of ascertaining diet], based on the identification

of opal phytoliths found bonded to the enamel surfaces of the teeth of extinct species, allows for identification of the actual plant remains eaten by an animal prior to its death. Thus the vegetative dietary preferences of an extinct species no longer have to be inferred but can be demonstrated directly through the identification of phytoliths, the inorganic remains of plant cells, on the teeth of extinct species.(Russell L. Ciochon, Dolores R. Piperno, and Robert G. Thompson, 1990)

In an analysis of 4 *Gigantopithecus* teeth, Ciochon et al. (2) (1990) identified 30 structures which were "indisputably phytoliths" on two of the teeth. These thirty broke down into two categories: the vegetative parts of grasses, and the fruits and seeds of dicotyledons.

Prior to the phytolith study Ciochon was pursuing a theory of massive bamboo consumption on the part of *Gigantopithecus* using analogy to the penchant of other megaherbivores to depend upon a single or limited number of plants. Creatures the size of *Gigantopithecus* would need a source which existed in abundance. The most likely candidate is bamboo. Further, the teeth seemed to point in that direction as well:

> The molar teeth of *Giganto* are low-crowned and flat, with very thick enamel caps. The premolars are molarized: that is, they have become broad and flattened, and thus resemble molars. The canine teeth are not sharp and pointed, but are rather broad and flat, more like what one would expect premolars to be; the incisors are small, peglike, and closely packed. These observations, combined with the massive jaw morphology, make it really an inevitable conclusion that the animal was adapted to the consumption of tough fibrous foods by cutting, crushing, and grinding

them.(Ciochon et al., 1990)

Ciochon et al., (1990) then go on to compare this morphology with that of the giant panda, another bamboo eater, and infers a diet of bamboo for *Gigantopithecus*.

While bamboo is a grass, the phytolith analysis does not technically either confirm or deny this theory, since it is not capable of defining the type of grass the phytolith came from. What was surprising to Ciochon was the suggestion of fruit in the diet of *Gigantopithecus*. Ciochon et al (2) (1990) have identified the fruit as belonging to a species in the family Moraceae or a closely related family and state "Judging from the present frequency of dental phytoliths in *Gigantopithecus*, fruits may have constituted a significant portion of the diet," and go on to note that the high sugar content of this type of fruit may be responsible for the high incidence of cavities in *Gigantopithecus* teeth (11%).

The results of this study are reported in less complete and less technical terms in the book *Other Origins* (Ciochon et al, 1990), and in a review of that book Jeffrey H. Schwartz (1991), which notes that a great deal is being drawn from the analysis of four teeth, upon only two of which were found phytoliths, with the greatest concentration on only one. Clearly a larger sample of teeth need to be similarly analyzed, but reading the report it is difficult not to share Ciochon's (et al. (2) 1990) excitement at the findings and for the employment of this technique in paleoanthropology in general.

Extinction
Ciochon (et al. 1990) propose three factors as being potentially related to the extinction of *Gigantopithecus blacki* and all are interrelated: dependence on bamboo, the giant panda, and *Homo erectus*. Bamboo is prone to periodic die offs, the exact reason for which is unknown. The giant panda was contemporaneous with *Gigantopithecus blacki* and may have been in competition with it for the same food source. The final straw, however, may have been the introduction of *Homo erectus* into the region. All three creatures, panda, *Giganto*, and *Homo*, may have been fond of

the sprouts of the bamboo as a food source (as are living pandas), which means that plants would have been consumed before they had a chance to reach maturity and reproduce. Further, *Homo erectus* may have been using bamboo for tools. In archaeology it was traditionally assumed that Asia was a cultural backwater during the stone age due to its lack of sophisticated stone tool kits like those found in Europe, but this attitude is changing as consideration is given to the wide variety of uses of bamboo, not only in theory, but as witnessed in practice in Asia through historical times into the present. Likewise, there is much debate around *Homo erectus'* proclivity for hunting, but another possible factor in the extinction of *Gigantopithecus blacki* is that it may have been hunted. Ciochon (et al., 1990) believes that it was likely a combination of factors, with the entry of *Homo erectus* into *Gigantopithecus'* range upsetting an already delicate balance. No one factor was likely absolute. For example, if *Homo erectus* had monopolized the fruit supply, it would have left *Gigantopithecus blacki* with no back up when a periodic bamboo die off occurred. This coupled with competition from the giant panda and sporadic hunting could have been enough to reduce breeding populations of *Gigantopithecus* below viable levels. (Ciochon et al., 1990)

The Myths
Some suggest that *Gigantopithecus blacki* did not in fact become extinct, and continues to exist as the Sasquatch and the Yeti. *Gigantopithecus blacki* could have crossed the Bering Land Bridge, the same way humans are thought to have entered the New World (Geoffrey Bourne, 1975, cited in Ciochon et al., 1990). So far, though there have been many alleged sightings, no indisputable physical evidence has been recovered. One is led to suspect that the question of Sasquatch (and related entities) is more for comparative mythology, cultural anthropology, or psychology, since an actual creature the size of *Gigantopithecus blacki* existing in numbers sufficient to qualify as a breeding population would not only leave physical remains, but would have an observable effect on their environment.

An old Sherpa once observed: "There is a yeti in the back of everyone's mind; only the blessed are not haunted by it." —Lama Surya Das, A Yeti Tale

Conclusion
We have cast the dragon's teeth, and something has sprung up. Is it a giant with the pleasing features of an orangutan and the impressive body of a gorilla? Perhaps it is a mega-gorilla, a prototype King Kong. Perhaps it will turn out to be something really surprising. One thing, though, is clear.

We need more data.

Appendix G:
Meganthropus:
Giant Man from Old Java

Early thoughts after discovery

Besides *Gigantopithecus*, there exists in the fossils of Asia, another type that has been problematic for paleoanthropologists. The series of finds in Java that have been grouped under the umbrella genus, *Meganthropus*, remain as elusive in their affinities as *Gigantopithecus*. They are, indeed, evidence of giants in Asia, and modern anthropologists appear to be uncomfortable with the possibility that they might be anything other than some large form of *Homo erectus*. But from the time of their discovery (when the term *Pithecanthropus* was used for what we call *Homo erectus* today), the giant *Meganthropus* were seen as something extraordinary.

The American Museum of Natural History anthropologist Franz Weidenreich extensively studied the fossil remains of large hominoids found in Asia. In his 1946 book, *Apes, Giants, and Man*, in a chapter entitled "Giants as Earliest Ancestors," Weidenreich wrote about these specimens, which little understood even today:

> Early in 1941, I received a letter from [anthropologist G. H. R.] von Koenigswald in which he announced the discovery of a fragment of another lower jaw, collected at the same site

as the jaw found earlier (Sangiran). But this time the critical teeth were still in their place and showed only slight attrition. A sketch of the piece was added. Von Koenigswald wrote that its proportions are enormous. I asked for a cast. It arrived just a couple weeks after Pearl Harbor. It could be gathered from the label that von Koenigswald intended to give the new human type, represented by this gigantic jaw, the name *Meganthropus paleojavanicus*, which means "giant man from old Java," and that he regarded the fragment as that of a male individual, which the fragment earlier, not yet recognized, was attributed by him to a female individual of the same type.

Later in the chapter, Weidenreich would reflect on *Meganthropus*, attempting to place it in some framework of fossils known in the 1940s:

The only skull bone which challenges the Java jaw in massiveness is the jaw of [Robert] Broom's *Paranthropus robustus* from southern Africa. This jaw belongs to that strange group of Australopithecinae which shows the typical organization of anthropoids mixed with some human features. The species name, *robustus*, was given by Broom because of this extraordinary appearance of the jaw.

The robustness of the *Pithecanthropus* skull is the link connecting it with the giant jaw from Java, in addition to the agreement in primitive human traits shown by both. This suggests that there was a continuous line of gigantic and nearly gigantic human forms characterized by a gradual reduction in size, this reduction going hand in hand with a progressive trend in other features.

For this reason I distinguish between the big *Pithecanthropus* skull of 1938 and the two earlier known, smaller skulls of *Pithecanthropus*....The big skull apparently represents a special type already on the way to giantism; therefore I gave it the name "*Pithecanthropus robustus.*"

Franz Weidenreich made a comparable observation worthy of repeating: "The molars of *Gigantopithecus* are more than one-third larger than those of *Meganthropus*, the Java giant, and almost twice as large as those of the big *Pithecanthropus*."

In G. H. R. von Koenigswald's own book on the topic, *The Evolution of Man* (Ann Arbor: University of Michigan Press, 1962: 74), he writes:

Meganthropus had the largest human lower jaw so far discovered. It is roughly the size of a gorilla's (length of first molar: *Meganthropus* 14.5 mm; gorilla 15.5 mm; average *Homo sapiens* 11.1 mm). [John] Robinson, who considers the jaw as that of *Paranthropus*, has shown that the two are comparable in size and structure.

Modern considerations

Meganthropus is a name commonly given to several large jaw and skull fragments from Sangiran, Central Java. The original scientific name was *Meganthropus palaeojavanicus*, and while it is commonly considered invalid today, the genus name has survived as something of an informal nickname for the fossil. As of 2005, the taxonomy and phylogeny for the specimens are still uncertain, although most paleoanthropologists consider them related to *Homo erectus* in some fashion. However, the names *Homo palaeojavanicus* and even *Australopithecus palaeojavanicus* are sometimes used as well, indicating the classification remains uncertain. Of particular interest is that the finds were sometimes regarded as those of giants, although that is technically unsubstantiated.

After the discovery of a robust skull in Swartkrans in 1948 (SK48), the name *Meganthropus africanus* was briefly applied. However, that specimen is now formally known as *Paranthropus robustus* and the earlier name is a junior synonym.

Some of these finds were accompanied by evidence of tool use similar to that of *Homo erectus*. This is the reason it is often linked with that species (although, of course, the tools found might not have been used by *Meganthropus*, but could have been used on them).

Fossil discoveries
The number of fossil finds has been relatively small, and it is a distinct possibility that they are a paraphyletic assemblage.

Meganthropus A/Sangiran 6
This large jaw fragment was first found in 1941 by von Koenigswald. Koenigswald was captured by the Japanese in World War II but managed to send a cast of the jaw to Franz Weidenreich, as noted above. Weidenreich described and named the specimen in 1945 and was struck by its size; it was the largest hominid jaw then known. The jaw was roughly the same height as a gorilla's but much thicker. Weidenreich considered acromegalic gigantism but ruled it out for not having typical features such as an exaggerated chin and small teeth compared to the jaw's size. Weidenreich never made a direct size estimate of the hominid it came from but said it was 2/3 the size of *Gigantopithecus*, which was twice as large as a gorilla; that would make the hominid somewhere around 8 feet (2.44 m) tall. The jawbone was used in part of Grover Krantz's skull reconstruction, which was 8.5 inches (21 centimeters) tall.

Meganthropus B/Sangiran 8
This jaw fragment was also described by Marks in 1953. It was around the same size and shape as the original mandible, but it was also severely damaged. Recent work by a Japanese/Indonesian team repaired the fossil, which was an adult, and showed it to be smaller than known specimens of *H. erectus*. Curiously, the

specimen retained several traits unique to the first mandibular find and not known in *H. erectus*. No size estimates have been made yet.

Meganthropus C/Sangiran 33/BK 7905

This mandibular fragment was discovered in 1979 and has some characteristics in common with previous mandible finds. Its connection with *Meganthropus* appears to be the most tenuous out of the mandibular discoveries.

Meganthropus D

This mandible and ramus was acquired by Sartono in 1993 and has been dated to between 1.4 and 0.9 million years ago. The ramus portion is badly damaged, but the mandible fragment appears relatively unharmed, although details of the teeth have been lost. It is slightly smaller than *Meganthropus* A and very similar in shape. Sartono, Tyler, and Krantz agreed that *Meganthropus* A and D were very likely to be representations of the same species, whatever it turns out to be.

Meganthropus I/Sangiran 27

Tyler described this specimen as being a nearly complete but crushed cranium within the size limit of *Meganthropus* and outside the (assumed) limit of *H. erectus*. The specimen was unusual for having a double temporal ridges, which almost meet at the top of the cranium, and a heavily thickened nuchal ridge.

Meganthropus II/Sangiran 31

This skull fragment was first described by Sartono in 1982. Tyler's analysis came to the conclusion that it was out of the normal range of *H. erectus*. The cranium was thicker, lower vaulted, and wider than any specimen previously recovered. It had the same double sagittal crest or double temporal ridge with a cranial capacity of around 800-1000cc. Since its presentation at the AAPA meeting in 1993, Tyler's reconstruction of Sangiran 31 has been accepted by most authorities. As with most fossils it was heavily damaged, but given the completeness of the post facial cranium, the chances of error in its reconstruction are very small. Tyler's accepted reconstruction of Sangiran 31 shows a double temporal ridge. The temporal muscles extend to the top of the parietal where they almost join. There are no other *Homo erectus* specimens that exhibits this trait. Krantz's reconstruction of making Sangiran 31 a giant *Homo habilis* is regarded as dubious by the scientific mainstream today.

Meganthropus III
  Another fossil with only tenuous ties to *Meganthropus*.

Scientific interpretation
Weidenreich theorized that *Meganthropus* was a descendant of *Gigantopithecus* that first gave rise to *Pithecanthropus* and then modern Asians. This hypothesis, part of the multi-regional theory of human evolution, has been discarded by mainstream paleoanthropology.

The second major theory, first proposed by J.T. Robinson, was that the *Meganthropus* finds are representative of a Southeast Asian australopithecine. This position has been adopted by several authorities, such as von Koenigswald and Krantz, but they are still regarded as a vocal minority. There was also discussion as to whether the finds are closer to *Australopithecus* or *Paranthropus*.

The majority of paleoanthropologists believe that *Meganthropus* is related to *H. erectus*, but it is not agreed upon just how closely they are related. Sartono believed that while it is related to *H. erectus*, the finds represent a new species, *H. paleojavanicus*. On the other side, several authors believe that they are merely the males of *H. erectus*, the alleged large size and robusticity being only due to early assumption that the females were males. There appears to be a consensus that there are some differences between *Meganthropus* and conventional *H. erectus*, but opinion is variable as to what the differences mean.

Another kind of True Giant?
Meganthropus has been the focus of numerous theories, none of which are supported by peer-reviewed authors. One popular

source, Polly Jae Lee's *Giant: The Pictorial History of the Human Colossus* (1970), promotes *Meganthropus* as a candidate for unknown hairy giants greater then 8 feet tall reported worldwide. The most common claim is that *Meganthropus* was a giant, with one unsourced claim putting them at 9 feet (2.75 m) tall and weighing 750 to 1000 pounds (340 to 450 kilograms). No exact height has been published in a peer reviewed journal recently, and none give an indication of *Meganthropus* being substantially larger than *H. erectus*.

There have been some rumors of post-cranial material, but those have either yet to be published or belong to *H. erectus*. Reports, apparently from Australian researcher Rex Gilroy, place *Meganthropus* in Australia, and attach it to giant tools and even modern day reports. However, almost all paleoanthropologists maintain that *Meganthropus* is only known from central Java. In a similar way, some Sasquatch researchers claim that Bigfoot is a modern *Meganthropus* (*Paranthropus*). But the Giant Man of Ancient Java may, after all, be a True Giant of some type, as yet unclassified.

In 2007, professors John Hawks and Cameron McCormick speculated in their weblogs that with the discovery of *Homo floresiensis* (the Hobbits), the issue of island gigantism might return as a focus of discussion in paleoanthropology. And with that attention in the next few years, there might be a return to debating *Meganthropus* again. Perhaps there is a new tomorrow for *Meganthropus paleojavanicus*, after all.

SOURCES

Introduction
1. "'X-Woman' a New Branch of Human Family Tree?" *Fox News*, March 24, 2010.
2. "Fossil DNA analysis may have revealed new human species," *USA Today*, March 25, 2010.
3. Loxton, Daniel. Personal correspondence with Loren Coleman. August 14, 2009

Chapter 1: The Universal Giant
1. Walter Stephens, *Giants in Those Days* (Lincoln: University of Nebraska Press, 1989: 2)

Chapter 2: The Earliest True Giant was *Gigantopithecus*
1. Franz Weidenreich, *Apes, Giants and Man* (Chicago: University of Chicago Press, 1946), 57-61; Michael Day, *Guide to Fossil Man* 1st ed. (New York: World, 1965), 262-7; Elwyn Simons and Peter Ettel, "Gigantopithecus," *Scientific American* 222(1): 76-85; and Robert Eckhardt, "Population Genetics and Human Origins," *Scientific American* 226 (1):94-103. Summary treatments of *Gigantopithecus* appear in such books as David Pilbeam, *The Ascent of Man* (New York: Macmillan, 1972), 86-89 and Frank E. Poirier, *Fossil Man: An Evolutionary Journey* (St. Louis: Mosby, 1973), 46-48.
2. G. H. Ralph von Koenigswald, *Meeting Prehistoric Man* (New York: Harper, 1956), 59.
3. Franz Weidenreich, *Apes, Giants and Man* (Chicago: University of Chicago Press, 1946).
4. Russell L. Ciochon, "The Ape That Was," *Natural History*

11/91, pp.54-63.
5. Simons and Ettel, "Gigantopithecus," 81-82.
6. John A. MacCulloch, "Eddic Mythology" in Vol. II of L. H. Gray, ed., *Mythology of All Races* (Boston: 1916), 275-84.
7. John Grant, *An Introduction to Viking Mythology* (Secaucus, New Jersey: Chartwell, 1990)

Chapter 3: True Giants in Europe
1. Lewis Spence, *The Minor Traditions of British Mythology* (London: Rider,1948) 58-86.
2. Ibid. 85.
3. Donald A. Mackenzie, *Teutonic Myth and Legend* (New York: William Wise,1934).
4. "The Abominable (Ballachulish) Snowman," *Sunday Post* (Scotland), 5 April 1959.
5. Spence, *Minor Traditions,* 101: "Cairngorms" in *Standard Encyclopedia of the World's Mountains* edited by Anthony Huxley (New York: Putnam, 1962), 99-100; Andrew Rothovius, "Big Gray Man of Ben MacDhui," *Fate* November 1980, 63-68. Affieck Gray, *Big Grey Man of Ben MacDhui* (Aberdeen: Impulse Books, 1970).
6. Janet Bord, "Some Scottish Monsters," *Flying Saucer Review* Vol. 18 No. 5 (Sept/Oct 1972), pp.16-17.
7. Doddy Hay, "The Grey Man of Braeriach," *Fifty Strange Stories of the Supernatural* (New York: Bell, 1974), 299-312.
8. Russell Beach, ed., *AA Touring Guide to Scotland* rev. ed. (Basingstoke, Hampshire: Fanum House, 1981), 93-94.
9. Grant, *An Introduction to Viking Mythology*; J. A. MacCulloch, "Eddic Mythology."
10. Jan Machal, "Slavic Mythology" in Vol. III of *The Mythology of All Races* (Boston, 1916).
11. Ivan T. Sanderson, "The Wudewasa or Hairy Primitives of Ancient Europe," *Genus* (Rome), 23(1-2): 109-40 (1967), Ivan T. Sanderson, *"Things"* (New York: Pyramid, 1967) 107-21: Richard Bemheimer, *Wild Men in The Middle Ages* (New York: Octagon Books, 1970).
12. *Encyclopedia Britannica* 1968 ed., Vol. 2, p.99; For discussions

of European "wild men" see the book *Wild Men of the Middle Ages* by Richard Bernheimer and the discussion of "Woodwoses" by Ivan Sanderson.
13. Ibid., Vol. 10, pp. 393-4.
14. John C. Lawson, *Modern Greek Folklore and Ancient Greek Religion: A Study in Survivals* (Cambridge: University Press, 1910), 190-255.
15. Tim Severin, *The Ulysses Voyage* (New York: Dutton, 1987), 88-98.

Chapter 4: The Asian Mountains Shelter Giants
1. William Smith, *Smith's Bible Dictionary* (Old Tappen, New Jersey: Fleming Revell, 1967), 34-35.
2. Adrienne Mayor, "Giants in Ancient Warfare," *Military History Quarterly*, 11 (Winter 1999), 98-101.
3. Mardiros H. Ananikian, "Armenian Mythology," in Vol. VII of L. H. Gray, ed. *Mythology of All Races* (Boston: 1916), 85-86.
4. Edmund Hillary and Desmond Doig, *High in the Thin Cold Air* (New York: Doubleday, 1961).
5. Ralph Izzard, *An Innocent on Everest* (New York: Dutton, 1951) 228.
6. Charles Stonor, *The Sherpa and the Snowman* (London: Hollis & Carter, 1955).
7. Bernard Heuvelmans, *On the Track of Unknown Animals* (New York: Hill and Wang, 1959),132-3.
8, London *Daily Herald*, 19 July 1935, and *Doubt: The Fortean Society Magazine*, No. 5, p. 4.
9. Edward T. C. Werner, *Myths and Legends of China* (London: Harrap, 1922), 387.
10. Ivan T. Sanderson, *Abominable Snowmen: Legend Come to Life* (Philadelphia: Chilton, 1961), 316-7.
11. Igor Iaseda, "The Abominable Snowman in the Mountains of Tajikistan?" *Moscow News Weekly*, No. 46 - 2982, Nov 22-29,1981, p.10.
12. Myra Shackley, *Still Living? Yeti, Sasquatch* and the *Neanderthal Enigma* (London: Thames and Hudson, 1983).
13. Alexandra Bourtseva, "The Footprint in Chuketka," *Technical*

Journal for Youth #6 (reprinted in *Bigfoot Co-Op*, Whittier, California, Volume 4, October 1983).

Chapter 5: The Giants of Southeastern Asia
1. "That Footprint" *Sunday New Straits Times*, 12 Feb. 1961.
2. *New Straits Times* 12, 15 Feb 1961; Harold Stephens, "Abominable Snowman of Malaysia, "*Argosy*, August 1971.
3. Wendy Moore. ed., *West Malaysia and Singapore* (Lincolnwood, Illinois: Passport Books, 1993),160.
4. Ronald McKie, *The Company of Animals* (New York: Harcourt, Brace, 1965), 30, 196-7.
5. Hassoldt Davis, *The Land of the Eye* (New York: Henry Holt, 1940), 111.
6. Deutsche Presse Agentur dispatch from Kuala Lumpur, 13 Jan 1995.
7. London *Evening News*, 1 Sept 1958.
8. Guy Piazzini, *The Children of Lilith* (New York: Dutton, 1959, 1960).

Chapter 6: Rumble in the Rainforest: Enter Orang Dalam
1. Bernama News Service, "Bigfoot Reported in Johor," 24 December 2005.
2. *The Star*, "Bigfoot May Help Eco-Tourism," 30 December 2005.
3. *New Straits Times*,"Closing in on Bigfoot, Foreign Media Set to Descend on Johor," 6 January 2006.
4. Jan McGirk, *The Independent*, "Expedition Tracks Malaysian Yeti," 22 January 2006.
5. *Malaya Mail*, "Locals Look for Bigfoot," 23 January 2006.
6. *The Straits Times*, "Bigfoot Expedition Delayed," 28 January 2006.
7. Jan McGirk, *The Independent*, "Malaysia launches hunt for 'Bigfoot' apeman after sightings in rainforest," 30 January 2006.
8. Rachel Gotbaum, National Public Radio's "Living On Earth,""Is Big Foot back? Maybe in Malaysia," 3 February 2006.
9. *World*, "On The Prowl," 11 February 2006.
10. Loren Coleman, Cryptomundo, "Harold Stephens: 1970s

Malaysian Bigfoot Tracker," 11 February 2006.

11. Loren Coleman, Cryptomundo, "Talking with Harold Stephens, Part II," 15 February 2006.

12. Loren Coleman, Cryptomundo, "Malaysian Bigfoot: Footprint Photos & Drawing," 17 February 2006.

13. Bernama News Service, "Are Bigfoot Really Orangutans," 18 February 2006.

14. *The Star*, "Bigfoot Still In Doubt," 19 February 2006.

15. Loren Coleman, Cryptomundo, "Are There Three Kinds of Malaysian Bigfoot?" 20 February 2006.

16. *New Straits Times*, "Zoologist: They Are Orang Dalam," 26 February 2006.

17. *The Star*, "Bigfoot mania making big news," 12 March 2006.

18. *NewsAsia*, "Investigators keen to present Bigfoot's footprint cast to Johor government," 13 March 2006.

19. Rehman Rashid, *New Strait Times*, "They seek him here, they seek him there…" 20 March 2006.

20. Loren Coleman, Cryptomundo, "Jumping Johor Roundup," 25 March 2006.

21. Loren Coleman, Cryptomundo, "McGirk Exclusive: Tracks Like Giant Gorillas," 6 April 2006.

22. *New Straits Times*, "Perhilitan Denies Catching Young Bigfoot," 19 April 2006.

23. Loren Coleman, Cryptomundo, "Orang Dalam (Harold Stephens Interview, Part III)," 3 August 2006.

Chapter 7: Tumble In The Jungle: Enter The Hoaxers

1. Peter Loh, Cryptomundo, "Malaysian Bigfoot = Surviving Homo Erectus?" 4 May 2006.

2. Loren Coleman, Cryptomundo, "The Face of the Johor Mawas," 6 May 2006.

3. R. Sittamparam, *New Straits Times*, "Book Has Bigfoot Pictures," 7 May 2006.

4. Bernama Wire Service, "Two Books Containing Pictures Of 'Bigfoot' Being Written," 13 May 2006.

5. Loren Coleman, Cryptomundo, "Johor Mawas Photo Analysis," 15 May 2006.

6. *The Star*, "Teams to Seek Proof of Bigfoot," 26 May 2006.

7. R. Sittamparam, *New Straits Times*, "Bigfoot: Johor to verify new info," 12 June 2006.

8. Zalina Mohd Som, *New Straits Times*, "Wild time in the jungle," 20 June 2006.

9. Loren Coleman, Cryptomundo, "New Name Nominee: Johor Hominid," 2 July 2006.

10. *China News Network*, "Malaysian researchers new discovery: Johor Bigfoot is the Peking Man," 3 July 2006.

11. Loren Coleman, Cryptomundo, "Johor Hominid: What Do You Look Like?" 7 July 2006.

12. Loren Coleman, Cryptomundo, "Johor Hominid Face?" 10 July 2006.

13. Nelson Benjamin, *The Star*, "Big Draw for Bigfoot Site," 13 July 2006.

14. Loren Coleman, Cryptomundo, "Johor Photo Release," 3 August 2006.

15. Loren Coleman, Cryptomundo, "Johor Hominid Photos: Hoax!" 4 August 2006.

16. Loren Coleman, Cryptomundo, "Johor Pix Hoax: More," 4 August 2006.

17. Loren Coleman, Cryptomundo, "More Johor Fallout: Peter Loh Reacts," 4 August 2006.

18. Loren Coleman, Cryptomundo, "Retiring 'Johor Hominid,'" 4 August 2006.

19. Loren Coleman, Cryptomundo, "Revisionist History: 'johor hominid,'" 5 August 2006.

20. Loren Coleman, Cryptomundo, "'A Species Odyssey,'" 8 August 2006.

21. Sandra Leong, *Singapore Straits Times*, "Bigfoot Hoax Exposed," 13 August 2006.

22. Loren Coleman, Cryptomundo, "Malaysian Bigfoot Search Continues," 14 August 2006.

Chapter 8: The Oceanic Adventures of True Giants

1. "Reward for a Waab," *Western Folklore* 9:387-8 (1950).

2. Matthew Green, "Volcanic eruption adds new twist to Comoros

legend," Reuters dispatch, 25 April 2005.
3. Frank Chapin Bray, *The World of Myths* (New York: Crowell, 1935), 232.
4. Loren Coleman and Patrick Huyghe, *Field Guide to Bigfoot and Other Mystery Primates* (San Antonio: Anomalist Books, 2006). 144-5.
5. Robert Quinton, *The Strange Adventures of Captain Quinton* (New York: Christian Herald, 1912), 282.
6. F. W. Christian, *The Caroline Islands* (1899) reprinted in *Ponape* ed. by Sibley S. Morrill (San Francisco: Cadleon Press, 1970), 126.
7. http://www.backyardoahu.com/index.html?olom3.htm
8. http://www.bigfootencounters.com/sbs.aikanaka.htm
9. Mary Kawena Pukui and Samuel Elbert, *The Hawaiian Dictionary* (Honolulu: University of Hawaii Press, 1971), 9.
10. Marius Boirayon, *Solomon Islands Mysteries*, Adventures Unlimited Press, 2010: 29-30.
11. Garcilaso de la Vega, *The Incas: The Royal Commentaries of the Inca* transl. Maria Jolas, ed. Alain Gheerbrant (New York: Avon Books, 1961), 327-8.

Chapter 9: Eastern True Giants
1. James Mooney, "Myths of the Cherokee," 19th Annual Report, Bureau of American Ethnology, No. 81, No. 106.
2. Robert F. Greenlee, "Folktales of the Florida Seminole," *Journal of American Folklore* 58:140 (1945).
3. H. P. Biggar, ed. *The Works of Samuel de Champlain* vol 1 (Toronto: The Champlain Society, 1922), 186-7.
4. Page 35 of Loren Coleman and Mark A. Hall, "From Atshen to Giants in North America," in *Sasquatch and Other Unknown Hominoids* ed. by V. Markotic and G. Krantz (Calgary: Western, 1984), 31-43.
5. Marvin A. Rapp, "Legend of the Stone Giants, " *New York Folklore Quarterly* 12 (1956): 280–282.
6. Hartley Burr Alexander, "North American," In *Mythology of All Races*. L.H. Gray (ed.), X. (New York: Cooper Square, 1964).
7. J. A. Burgosse,"Windigo!," *Beaver*, March 1947,4-5.

8. Charles M. Skinner, *American Myths and* Legends vol. 1 (Philadelphia: Lippincott, 1900), 37-38.
9. James Owen Dorsey, "Siouan Folk-lore and Mythologic Notes," *The American Antiquarian*, 7:105-8 (1885).
10. Personal communication, Brian Rathjen, Editor of *Backroads Motorcycle TourMagazine* to Loren Coleman, October 23, 2010.
11. In Minnesota: "Misabe," list compiled April 10, 1907, Red Lake, p. 40. Rev. Joseph Gilfillan manuscripts at the Minnesota Historical Society. In Quebec: D. S. Davidson, "Folktales from Grand Lake Victoria, Quebec," *Journal of American Folklore* 41:275 (1928).
12. Alpheus Hyatt Verrill, *Along New England Shores* (New York: Putnam, 1936), 138.
13. Corinth (Mississippi) *Corinthian*, 16, 18 March 1976. Memphis (Tennessee) *Commercial Appeal* 23 March 1976.
14. Brookhaven (Mississippi) *Leader*, 16 February 1977.
15. Longview (Texas) *Journal*, 28 November 1976.
16. Reported to the Bigfoot Field Researchers Organization website in 1999.
17. Winnipeg (Manitoba) *Sun*, 17 July 1988.
18. *The Creature Research Journal*, Spring 1994 (Published by Paul G. Johnson, Versailles, Pennsylvania).

Chapter 10: The Giants in the American West
1. Christopher Roth email to Loren Coleman on 20 September 2003, posted at the Bigfoot Yahoo Groups, 20 September 2003.
2. Sidney Warren, *Farthest Frontier: The Pacific Northwest* (New York: Macmillan, 1949), 159-62.
3. Stephen Franklin, "The Sasquatch," *Weekend Magazine*, 4 April 1959; reprinted in *Fate* Magazine as "On the Trail of the Sasquatch," June 1960.
4. Margaret Ormsby, ed., *A Pioneer Gentlewoman in British Columbia* (Vancouver: University of British Columbia Press, 1976).
5. Clark B. Firestone, *Coasts of Illusion* (New York: Harper, 1924), 192.

6. Russell Annabel, "Long Hunter Alaskan Style," *Sports Afield* July 1963.
7. W. R. Abercrombie, "Report of a Supplementary Expedition into the Copper River Valley, 1884," in *Compilation of Narratives of Explorations in Alaska: 1869-1900* (Washington: Government Printing Office, 1900).
8. Franz Boas, "Traditions of the Tsetsaut," *Journal of American Folklore* 10: 44-47 (1897).
9. Michael H. Mason, *The Arctic Forests* (London: Hodder & Stoughton, 1924).
10. John J. Honigmann, "The Kaska Indians," Yale University Publications in Anthropology, 51, 1954, 103.
11. Ray Crowe, "Editorial," *The Track Record: Western Bigfoot Society Newsletter* (Portland, Oregon) No. 65 (March 1997), 5
12. Charles Edson, *My Travels with Bigfoot* (Los Angeles: Crescent, 1979), 13-14; Mark A. Hall, "Charles Edson's Quest for Bigfoot," *Wonders* 8(2):35-43 (June 2003).
13. John Green, *Year of the Sasquatch* (Agassiz, British Columbia: Cheam, 1970), 68-70.
14. Great Falls (Montana) *Tribune*, 26 August 1977, 3 September 1977.
15. Mick Gidley and Ruth Gidley, "Plateau and Basin" in *Native American Myths and Legends* ed. by Colin Taylor (London: Salamander Books, 1994), 61.
16. C. Hart Merriam, *The Dawn of the World: Myths and Tales of the Miwok Indians of California* (Cleveland: Arthur H. Clark, 1910; reprint by University of Nebraska Press, 1993).
17. Ryan Hall in Farmington (New Mexico) *Daily Times*, 31 August 2005.
18. Mark A. Hall, *The Yeti, Bigfoot & True Giants* 3d ed. (Wilmington, North Carolina: Mark A Hall Publications, 2005), 86, 88.

Chapter 11: The World of True Giants

1. Firestone, *Coasts of Illusion*, 192.
2. Merriam, *Dawn of the World*, 170.
3. "Reports tell of Canadian Monster Men," United Press dispatch

in the Hammond (Indiana) *Times*, 25 October 1935. (In *Big News Prints*, compiled by Scott McClean, 2005, p. 49.)

Chapter 12: The Future of True Giants
1. George and Helen Papashvily, *Anything Can Happen* (New York: Harper & Brothers, 1945).
2. George and Helen Papashvily, *Home, And Home Again* (New York: Harper & Row, 1973).
3. Ibid, 128.
4. Ibid, 130.
5. Merriam, *Dawn of the World.*

Appendix C: Giant Bones
1. Gordon Strasenburgh Jr., "Perceptions and Images of the Wild Man" in the *Scientist Looks at the Sasquatch* edited by R. Sprague and G. Krantz (Moscow, ID: University Press of Idaho, 1977) 125.
2. Charles Hapgood, *Path of the Pole* (Philadelphia: Chilton, 1970).
3. John Green, *On the Track of the Sasquatch* (Agassiz, B.C.: Cheam, 1969), 72.
4. Stephen Franklin, "On the Trail of the Sasquatch," Canada Wide Feature Service, reprinted in *Fate Magazine*, June 1960, pp. 54-63.
5. Dana and Ginger Lamb, *Quest for the Lost City* (1951; reprint Santa Barbara California: Santa Barbara Press,1984), 64-65.
6. B. Ann Slate and Alan Berry, *Bigfoot* (NY: Bantam Books, 1976), 160-65.
7. Matt Moneymaker, "Buried treasure in Chatsworth," *Bigfoot Co-Op* (Whittier, CA), December 1993.
8. Hall, "The Real Bigfoot," 122.
9. Robert L. Thomas, "Canyon Explorer Seeks Hardy Companion," *Arizona Republic*, 23 November 1974, B!, B2.
10. Jack Clayton, The Giants of Minnesota," *Doubt: the Fortean Society Magazine* No. 35, pp. 120-22 (the issue has no date; approx. 1951).
11. W. Mead Stapler, "A Mystery in History," *The North Jersey*

Highlander, Spring 1973, p.2.
12. "Giant Skeletons," *Pursuit,* July 1973, 69.
13. Willy Ley, *Exotic Zoology* (NY: Viking Press, 1959) 28-35.
14. Francis Buckland, *Curiosities of Natural History, Fourth Series* (London: Richard Bentley, 1891), 27-28. Reprinted by Cosimo Books, New York, New York, 2008, in the "Loren Coleman Presents" series.
15. Adolph Bandelier, "Traditions of Precolumbian Landings on the Western Coast of South America," *American Anthropologist* 7: 250-70 (1905).
16. Edinond Temple, *Travels in Various Parts of Peru* vol.2 (London, 1830), 320-22.

Appendix D: Giant Skulls
1. Hardy, Prof. Alister, quoted in AP verbatim report of March 6, 1960 conference of marine scientists, Brighton, England in *New York Herald Tribune* 7 March 1960.
2. Koenigswald, G. H. R. von, *Meeting Prehistoric Man,* New York: Harper & Bros., 1956.
3. "A Giant Ape of 500,000 Years Ago: New Light on the Monster Gigantopithecus of Prehistoric China," *Illustrated London News,* 13 April 1957.

Appendix E: The Toonijuk
Knud V. J. Rasmussen, *Reports of the Fifth Thule Expedition. 1921-1925, The Danish Expedition to Arctic North America* (Copenhagen: Glydendalske, Nordisk Forlag, 1927).
Katharine Scherman. *Spring on an Arctic Island* (Boston: Little, Brown and Company, 1956).

Appendix F: Teeth of the Dragon by Eric Pettifor
Ciochon, Russell L., Dolores R. Piperno, and Robert G. Thompson, 1990. "Opal phytoliths found on the teeth of the extinct ape *Gigantopithecus blacki*: Implications for paleodietary studies." *Proceedings of the National Academy of Science,* 87: 8120-8124.

Ciochon, Russel L., John Olsen, and Jamie James, 1990. *Other Origins: The Search for the Giant Ape in Human Prehistory.* New York: Bantam Books.

Corruccini, Robert S. 1973. "Multivariate Analysis of *Gigantopithecus* Mandibles." *American Journal of Physical Anthropology,* 42: 167-170.

Frayer, David W. 1972. "*Gigantopithecus* and Its Relationship to *Australopithecus.*" *American Journal of Physical Anthropology,* 39: 413-426.

Hamilton, Edith 1942. *Mythology.* Boston: Little Brown and Company.

Schwartz, Jeffry H. 1991. "Book Review of *Other Origins: The Search for the Giant Ape In Human Origins.*" *American Anthropologist,* 93: 1029-1030.

Simons, Elwyn L., and Peter C. Ettel 1970. *Gigantopithecus. Scientific American,* January, 1970: 77-85.

Von Koenigswald, G.H.R. 1952. "*Gigantopithecus blacki* Von Koenigswald, a giant fossil hominoid from the pleistocene of southern China." *Anthropological Papers of the American Museum of Natural History,* 43: 295-325

Appendix G: *Meganthropus*: Giant Man From Old Java

Ciochon, Russell, John Olsenm and Jamie James, *Other Origins: The Search for the Giant Ape in Human Prehistory.* New York: Bantam Books, 1990.

Coleman, Loren, and Patrick Huyghe. *The Field Guide to Primates and Other Mystery Primates,* New York: Anomalist Books, 2006.

Durband, A. C. "A re-examination of purported Meganthropus cranial fragments." Paper not yet published. Abstract available in the *American Journal of Physical Anthropology* supplements for 2003.

Heuvelmans, Bernard. *On the Track of Unknown Animals.* London: Rupert Hart Davis, 1962.

Kaifu, Yousuke, Fachroel Aziz, and Hisao Baba. "Hominid Mandibular Remains From Sangiran: 1952-1986 collection." *American Journal of Physical Anthropology,* 2005.

Kramer, A. "A Critical Analysis of Southeast Asian Australopithecines." *Journal of Human Evolution* Volume 26, number 1, 1994.

Krantz, Grover, S. Sartono, and D. Tyler. "A New Meganthropus Mandible from Java." *Human Evolution*, 1995.

Lee, Polly Jae, *Giant: The Pictorial History of the Human Colossus*, A. S. Barnes, South Brunswick: 1970.

Tyler, D. "Taxonomic Status of 'Meganthropus' Cranial Material." Supplements of the *American Journal of Physical Anthropology*, 1993.

von Koenigswald, G. H. R. *The Evolution of Man*. Ann Arbor: University of Michigan Press, 1962.

Weidenreich, Franz. *Apes, Giants, and Men*. Chicago: University of Chicago Press, 1946.

ABOUT THE AUTHORS

Mark A. Hall

A FORTEAN, CRYPTOZOOLOGIST, author, and theorist, Mark A. Hall was born on June 14, 1946, raised in the heartland of America in Bloomington, Minnesota, educated in anthropology, and served as a linguist in West Berlin in the midst of the Cold War. Besides being an editor at an archeological society in Minnesota after his military service, Hall has worked in human relations in various branches of the federal government while in his home state. Hall has been intrigued by nature's anomalies for most of his life. For 50 years, he has actively pursued historical records and eyewitness testimony concerning cryptozoological phenomena. He has traveled extensively throughout the Americas and the world. Hall was involved with Ivan T. Sanderson in investigations of the Minnesota Iceman, appeared on "Unsolved Mysteries" detailing that involvement, and was a director of the Sanderson-founded Society for the Investigation of the Unexplained (SITU) in the early 1970s.

For years, he edited and published the journal *Wonders*, devoted mostly to cryptozoology and to other Forteana. Hall's early books on natural mysteries and unknown hairy hominoids were self-published. His most recent book, *Thunderbirds: America's Living Legends of Giant Birds*, is published by Paraview Press, and his *Merbeings: The True History of Mermaids and Lizardmen* is forthcoming from Anomalist Books.

Hall's theories, his challenges even within cryptozoology, his memory for details, and his intellectual insights have always impressed his colleagues.

Ivan T. Sanderson introduced Loren Coleman to Mark Hall in 1969, and they have been correspondents, research associates, and friends for over 40 years.

Loren Coleman

LOREN COLEMAN WAS born on July 12, 1947 in Virginia, raised in Decatur, Illinois where he began his cryptozoology research in March 1960, briefly lived in northern California, and has lived in New England since 1975.

Coleman is acknowledged as the world's leading living cryptozoologist, and the U.S.' most grounded, skeptically open-minded Bigfoot researcher. The founder of the International Cryptozoology Museum in 2003, he frequently appears on documentaries on Animal Planet, Discovery, History, A&E, and SyFy channels. His consultant roles have involved working with "MonsterQuest," "Lost Tapes," "Unsolved Mysteries," and on *The Mothman Prophecies* movie. He is a frequent guest on "Coast to Coast AM." Coleman's 30-plus books, including *The Field Guide to Bigfoot Worldwide* and *The Field Guide to Lake Monsters, Sea Serpents, and Other Mystery Denizens of the Deep* (both with Patrick Huyghe), *Mysterious America, Bigfoot! The True Story of Apes in America*, and *Cryptozoology A to Z* (with Jerome Clark), are high-mark standards within the field.

Obtaining an undergraduate degree from Southern Illinois University at Carbondale, Coleman majored in anthropology, minored in zoology, and did some summer work in archaeology. He received a graduate degree in psychiatric social work from Simmons College in Boston, and did doctorate work in both social anthropology at Brandies University and sociology at the University of New Hampshire's Anthropology/Sociology Department without obtaining a Ph.D. He has taught documentary film, social work, cryptozoology, and anthropology courses for more than 20 years at six different New England universities.

He lives in Portland, Maine, with his partner, professor Donna Bird, and gives tours of his public museum when he's not on fact-finding expeditions, visiting his sons Malcolm and Caleb in New England, and keeping in touch with family members and friends throughout the U.S. and the world. Infrequently, when time allows, he continues doing human mysteries research at Portland Sea Dogs and Red Sox baseball games.

CPSIA information can be obtained at www.ICGtesting.com
Printed in the USA
BVOW03s0142200913

331629BV00008B/133/P